Biomarkers for Traumatic Brain Injury

RSC Drug Discovery Series

Editor-in-Chief:
Professor David Thurston, *London School of Pharmacy, UK*

Series Editors:
Dr David Fox, *Pfizer Global Research and Development, Sandwich, UK*
Professor Salvatore Guccione, *University of Catania, Italy*
Professor Ana Martinez, *Instituto de Quimica Medica-CSIC, Spain*
Professor David Rotella, *Montclair State University, USA*

Advisor to the Board:
Professor Robin Ganellin, *University College London, UK*

Titles in the Series:

1: Metabolism, Pharmacokinetics and Toxicity of Functional Groups
2: Emerging Drugs and Targets for Alzheimer's Disease; Volume 1
3: Emerging Drugs and Targets for Alzheimer's Disease; Volume 2
4: Accounts in Drug Discovery
5: New Frontiers in Chemical Biology
6: Animal Models for Neurodegenerative Disease
7: Neurodegeneration
8: G Protein-Coupled Receptors
9: Pharmaceutical Process Development
10: Extracellular and Intracellular Signaling
11: New Synthetic Technologies in Medicinal Chemistry
12: New Horizons in Predictive Toxicology

13: Drug Design Strategies: Quantitative Approaches
14: Neglected Diseases and Drug Discovery
15: Biomedical Imaging
16: Pharmaceutical Salts and Cocrystals
17: Polyamine Drug Discovery
18: Proteinases as Drug Targets
19: Kinase Drug Discovery
20: Drug Design Strategies: Computational Techniques and Applications
21: Designing Multi-Target Drugs
22: Nanostructured Biomaterials for Overcoming Biological Barriers
23: Physico-Chemical and Computational Approaches to Drug Discovery
24: Biomarkers for Traumatic Brain Injury

How to obtain future titles on publication:
A standing order plan is available for this series. A standing order will bring delivery of each new volume immediately on publication.

For further information please contact:
Book Sales Department, Royal Society of Chemistry, Thomas Graham House, Science Park, Milton Road, Cambridge, CB4 0WF, UK
Telephone: +44 (0)1223 420066, Fax: +44 (0)1223 420247, Email: books@rsc.org
Visit our website at http://www.rsc.org/Shop/Books/

Biomarkers for Traumatic Brain Injury

Edited by

Svetlana A. Dambinova
Distinguished Professor, WellStar College of Health and Human Services, Kennesaw State University, GA, USA
E-mail: sdambino@kennesaw.edu

Ronald L. Hayes
Founder & President, Banyan Biomarkers, Inc.; Director, Banyan Laboratories, Alachua, FL, USA
E-mail: rhayes@banyanbio.com

Kevin K. W. Wang
Executive Director, Center of Neuroproteomics and Biomarkers Research, Departments of Psychiatry & Neuroscience, McKnight Brain Institute, University of Florida, Gainesville, FL, USA
E-mail: kawangwang17@gmail.com

RSC Publishing

RSC Drug Discovery Series No. 24

ISBN: 978-1-84973-389-2
ISSN: 2041-3203

A catalogue record for this book is available from the British Library

© Royal Society of Chemistry 2012

The RSC is not responsible for individual opinions expressed in this work.

Published by The Royal Society of Chemistry,
Thomas Graham House, Science Park, Milton Road,
Cambridge CB4 0WF, UK

Registered Charity Number 207890

For further information see our web site at www.rsc.org

Printed in the United Kingdom by Henry Ling Limited, at the Dorset Press, Dorchester, DT1 1HD

Foreword

A conservatively estimated 3–4 million people experience sport- and recreation-related concussions annually in the United States. Additionally, 300 soldiers per month experience traumatic brain injuries (TBI), making it the most common cause of disability in the military today. A substantial number experience intractable headaches, memory impairment, cognitive abnormalities, depression, and severe debilitation – the so-called post-concussion syndrome.

Presently, the diagnosis of mild TBI is based on history and clinical findings. Neuroimaging with CAT scans and MRI, in most cases, is normal. Biochemical markers of mild TBI are urgently needed to improve diagnosis, monitor recovery, and guide therapeutic interventions.

In this unique volume, internationally recognized authorities in the field of TBI and biomarkers summarize the most current literature in the pathogenesis of TBI, describe promising biomarkers for detecting even the most subtle effects of cerebral concussion, and discuss future trends and the profound clinical implications of this technology in the management of all with TBI.

Editors Dambinova, Hayes, and Wang, through their multidisciplinary approach, provide a volume that serves as a substrate of understanding for future trends in the management of TBI with biomarker assays – a welcome addition for basic scientists and clinicians alike.

Joseph C. Maroon, MD
Vice Chairman and Clinical Professor
Heindl Scholar in Neuroscience
Department of Neurosurgery
University of Pittsburgh Medical Center
Team Neurosurgeon, The Pittsburgh Steelers
Pittsburgh, PA, USA

RSC Drug Discovery Series No. 24
Biomarkers for Traumatic Brain Injury
Edited by Svetlana A. Dambinova, Ronald L. Hayes and Kevin K. W. Wang
© Royal Society of Chemistry 2012
Published by the Royal Society of Chemistry, www.rsc.org

Preface

Neuroscience researchers have made remarkable progress in understanding the molecular basis of mental and behavioral complexities of the brain. This work has, in turn, stimulated development of modern technologies for brain imaging under normal and pathological conditions. In this book, we present an integrative review of the state-of-the art approaches to utilize biochemical and neuroimaging markers to diagnose, monitor, and help manage traumatic brain injury (TBI).

Every year in the United States alone, an estimated 1.7 million cases of TBI occur, which translates to 4.7 new cases per 1000 people per year. Similar incident rates of TBI have also been reported in Europe and Asia. Current and novel prospective biomarkers for the assessment of TBI, particularly mild TBI, are examined using a multidisciplinary approach involving biochemistry, molecular biology, and clinical chemistry and interrelationships among nervous, vascular, and immune systems are highlighted. Recently, the medical application of biomarkers for TBI was discussed in the plenary session "Biomarkers and TBI" at the International Conference on Recent Advances in Neurotraumatology (ICRAN 2010, St. Petersburg, Russia, July 2010) and in presentations at the 6th International Conference on Biochemical Markers for Brain Damage (BMBD, Lund, Sweden, May 2011). In this book, outstanding scientists, young researchers, and prominent clinicians from academia and industry – including those who presented at the two international conferences – have contributed their latest data describing advanced proteomic and degradomic technologies in the development of novel biomarker assays.

The focus of this volume is on detection of brain/blood-based biomarkers to improve diagnostic certainty of mild TBI in conjunction with radiological and clinical findings. The book will enhance the knowledge of scientists, pharmacologists, chemists, medical students, and graduate students, and we

RSC Drug Discovery Series No. 24
Biomarkers for Traumatic Brain Injury
Edited by Svetlana A. Dambinova, Ronald L. Hayes and Kevin K. W. Wang
© Royal Society of Chemistry 2012
Published by the Royal Society of Chemistry, www.rsc.org

hope that the data presented on biomarkers will stimulate young researchers to study both diagnostic and novel approaches to the emergent treatment of mild TBI.

We express our deep gratitude to each author and to the editors at D. A. Hughes & Associates (Lusby, MD, USA) and the Royal Society of Chemistry (Cambridge, UK) for their contributions to this book. We wish to especially acknowledge the generous contributions of Dr Stefania Mondello, who edited several chapters of this book on which she is not an author. We also appreciate the efforts of Gwen Jones and Juliet Binns, RSC Publishing, for guiding production of this volume and in keeping us all on track to ensure its timely and successful completion.

The research of several US contributors has been supported by grants from US Army Medical Research and Materiel Command.

Svetlana Dambinova
Ronald L. Hayes
Kevin K. W. Wang

Contents

RSC Drug Discovery Series No. 24
Biomarkers for Traumatic Brain Injury
Edited by Svetlana A. Dambinova, Ronald L. Hayes and Kevin K. W. Wang
© Royal Society of Chemistry 2012
Published by the Royal Society of Chemistry, www.rsc.org

List of Contributors

Ramona Åstrand Department of Neurosurgery, Rigshospitalet, Copenhagen, Denmark

Arthur Bagumyan, MS GRACE Laboratories, LLC, Decatur, GA, USA

Randall R. Benson, MD Clinical Assistant Professor, Departments of Neurology and Radiology, Wayne State University School of Medicine, Detroit, MI, USA; Medical Director—Center for Neurological Studies, Novi, MI, USA

Kerstin Bettermann, PhD, MD Associate Professor, Department of Neurology, Penn State College of Medicine, Hershey, PA, USA

Russell L. Blaylock, MD Theoretical Neurosciences, LLC, Visiting Professor Biology, Belhaven University, Jackson, MS, USA

Svetlana A. Dambinova, DSc, PhD Distinguished Professor, Head of Laboratory of Brain Biomarkers, WellStar College of Health & Human Services, Kennesaw State University, Kennesaw, GA, USA

Uliana I. Danilenko, PhD CIS Biotech, Inc., Decatur, GA, USA

Christian Foerch, MD Professor, Department of Neurology, University Hospital, J.W. Goethe University, Frankfurt, Germany

Sarah Gill, MS, ATC, LAT Club Sports Athletic Trainer, KSU Club Sports Owls Nest, Kennesaw State University, Kennesaw, GA, USA

Olena Glushakova, MS Principal Scientist, Banyan Laboratories, Inc., Alachua, FL, USA

Hector Gutierrez, PhD Associate Professor, Department of Mechanical and Aerospace Engineering, Florida Institute of Technology, Melbourne, FL, USA

E. Mark Haacke, PhD Director, MR Research Facility, Professor of Biomedical Engineering and Radiology, Wayne State University School of Medicine, Detroit, MI, USA

RSC Drug Discovery Series No. 24
Biomarkers for Traumatic Brain Injury
Edited by Svetlana A. Dambinova, Ronald L. Hayes and Kevin K. W. Wang
© Royal Society of Chemistry 2012
Published by the Royal Society of Chemistry, www.rsc.org

Ronald L. Hayes, PhD   Center of Innovative Research, Banyan Biomarkers Inc., Alachua, FL, USA

Galina A. Izykenova, PhD   GRACE Laboratories, LLC, Decatur, GA, USA

German A. Khunteev, PhD, MD   CIS Biotech, Inc., Decatur, GA, USA

Daniel R. Kirk, PhD   Associate Professor, Department of Mechanical and Aerospace Engineering, Florida Institute of Technology, Melbourne, FL, USA

Zhifeng Kou, PhD   Assistant Professor, Departments of Biomedical Engineering and Radiology, Wayne State University School of Medicine, Detroit, MI, USA

Joseph C. Maroon, MD   Vice Chairman and Clinical Professor, Heindl Scholar in Neuroscience, Department of Neurosurgery, University of Pittsburgh Medical Center, Team Neurosurgeon, The Pittsburgh Steelers, Pittsburgh, PA, USA

Stefania Mondello, MD, PhD, MPH   Director of Clinical Research, Center of Innovative Research, Banyan Laboratories, Inc., Alachua, FL, USA; Research Fellow, Department of Anesthesiology, University of Florida, Gainesville, FL, USA

Miroslav M. Odinak, DSc, PhD, MD   Professor & Chair, Department of Neurology, Russian Medical Military Academy, St. Petersburg, Russia

Alexander V. Panov, PhD, MD   Research Fellow, Laboratory of Brain Biomarkers, WellStar College of Health & Human Services, Kennesaw, Kennesaw State University, Kennesaw, GA, USA

Victor Prima, PhD   Principal Scientist, Banyan Laboratories, Inc., Alachua, FL, USA

Bertil Romner, MD, PhD   Department of Neurosurgery, Rigshospitalet, Copenhagen, Denmark

Alexey V. Shikuev, MD   Research Fellow, Laboratory of Brain Biomarkers, WellStar College of Health & Human Services Kennesaw, Kennesaw State University, Kennesaw, GA, USA

Alexander A. Skoromets, Acad. RAMN, DSc, PhD, MD   Professor & Chair, Department of Neurology & Neurosurgery, Pavlov State Medical University, St. Petersburg, Russia

Taras A. Skoromets, DSc, PhD, MD   Professor, Department of Neurology & Neurosurgery, Pavlov State Medical University, St. Petersburg, Russia

Dmitrii I. Skulyabin, PhD, MD   Department of Neurology, Russian Medical Military Academy, St. Petersburg, Russia

Julia E. Slocomb, BS   Clinical Research Coordinator, Department of Neurology, Penn State College of Medicine, Hershey, PA, USA

Laura St. Onge, BA   Associate Director, Competitive Sports, KSU Sports and Recreation Park/Owls Nest, Kennesaw State University, Kennesaw, GA, USA

Artem Svetlov   Lab. Assistant, Banyan Laboratories, Inc., Alachua, FL, USA

Stanislav I. Svetlov, PhD, MD   Associate Director, Banyan Laboratories, Inc., Alachua, FL, USA; Adjunct Associate Professor, Department of Medicine, University of Florida, Gainesville, FL, USA

Richard L. Sowell, PhD, RN, FAAN Professor and Dean, WellStar College of Health & Human Services Kennesaw, Kennesaw State University, Kennesaw, GA, USA

Johan Undén, MD, PhD Department of Anesthesia and Intensive Care, Skåne University Hospital, Malmö, Sweden

Kevin K. W. Wang, PhD Executive Director, Center of Neuroproteomics and Biomarkers Research, Departments of Psychiatry & Neuroscience, McKnight Brain Institute, University of Florida, Gainesville, Florida, USA

CHAPTER 1
Clinical Relevance of Biomarkers for Traumatic Brain Injury

KERSTIN BETTERMANN* AND JULIA E. SLOCOMB

Penn State College of Medicine, Department of Neurology, 500 University Drive, Hershey, PA 17033, USA
*Email: kbettermann@hmc.psu.edu

1.1 Introduction

Traumatic brain injury (TBI) is defined as damage to the brain due to sudden trauma, either by penetrating or, more commonly, by closed head injury. TBI can be either focal or diffuse and is classified as mild, moderate, or severe, as defined in Table 1.1 by the U.S. Department of Defense and Veterans Administration Traumatic Brain Injury Task Force.

Moderate and severe TBI have long been known to lead to impairments in motor and vision function, cognition, attention, memory, and executive functions.[1] However, in recent years it has been discovered that even mild TBI (mTBI) can also have lasting and commonly disabling effects including headache, concentration difficulties, insomnia, mood disturbances, and dizziness.[2] Regardless of severity, TBI involves two phases – the primary injury, which is acquired during impact, and the secondary injury, which is due to brain damage that evolves throughout the subsequent hours and days.[3] Although some of the contributing factors of this secondary damage are understood, such

RSC Drug Discovery Series No. 24
Biomarkers for Traumatic Brain Injury
Edited by Svetlana A. Dambinova, Ronald L. Hayes and Kevin K. W. Wang
© Royal Society of Chemistry 2012
Published by the Royal Society of Chemistry, www.rsc.org

Table 1.1 TBI severity scale based on U.S. Department of Defense and Veterans Administration Traumatic Brain Injury Task Force recommendations. Multiple grading systems for TBI and concussion are available. This scale may be of use for biomarker trials, as it takes into account the Glasgow Coma Scale score as well as clinical and neuroimaging findings as reference standards.[13]

Mild	*Moderate*	*Severe*
Normal structural imaging	Normal or abnormal structural imaging	Normal or abnormal structural imaging
LOC = 0–30 min	LOC > 30 min to 24 h	LOC > 24 h
AOC = momentarily, up to 24 h	AOC > 24 h; severity based on other criteria	
PTA = 0–1 days	PTA > 1 to 7 days	PTA > 7 days
GCS = 13–15 points	GCS = 9–12 points	GCS = 3–8 points

AOC, alteration of consciousness/mental state; LOC, loss of consciousness; PTA, post-traumatic amnesia; GCS, Glasgow Coma Scale.

as cerebral hyperemia or dysfunction of cerebrovascular autoregulation, diagnosis and treatment of secondary brain injury remains rudimentary despite years of research. As mild TBI typically is not associated with any changes on brain MRI and is difficult to assess by standard diagnostic workup, its pathophysiology is not well known. There is no objective test for mild TBI, other than history and neurological examination, and there remains great disparity in our ability to identify mild TBI and predict clinical outcome. Therefore, there is an unmet need to improve the diagnosis of patients with mild TBI and to identify individuals who are susceptible to secondary brain injury. Recent advances in biomarker research may help to fill these gaps by improving diagnostic certainty. In the future, biomarkers of mild TBI may help to guide therapeutic decisions and could be useful in predicting clinical outcome.

1.2   Epidemiology and Health Care Economics

Globally, TBI is the leading cause of mortality and morbidity in individuals under the age of 45 years.[4] Approximately 1.7 million people in the United States sustain a TBI each year, resulting in more than 235 000 hospitalizations and 50 000 deaths.[5,6] The rate of death as a result of TBI is three times higher in males than in females. However, estimated morbidity and mortality rates for TBI are probably imprecise due to inaccurate recording of cause of death, cause of injury, circumstances of injury, and inconsistencies in the diagnosis of TBI.[6] A significant number of mild TBI cases probably go unrecorded, as many victims do not seek medical attention. The National Institute of Neurological Disorders and Stroke estimates that between 2.5 and 6.5 million Americans alive today have had one or multiple TBIs.[7] Mild TBI, which accounts for 80–90% of all cases, is the most prevalent form of brain injury. Among civilians, TBI is most common in children ages 0–4 years, adolescents age 15–19 years, and adults age 65 or older.[5]

U.S. military combat missions, which have increased in recent years, have created an additional large population of young TBI survivors, mostly due to blast injuries. As of 2006, roughly 30% of injured U.S. soldiers returning from Iraq to Walter Reed Army Medical Center were diagnosed with TBI.[8] While the precise number of TBIs among U.S. troops remains unknown, it has been estimated by the U.S. Department of Defense that 212 742 service men and women have been diagnosed with TBI between January 2000 and May 2011.[9] TBI is commonly referred to as the "signature injury" of the wars in Iraq and Afghanistan, as TBI is thought to be more prevalent now than in any past wars in U.S. history.[8]

Given that TBI is most common among young adults, costs to society can be significant. Depending on the severity of their injuries and disabilities, these survivors often require specialized care for the rest of their lives.[10] An estimated 5.3 million Americans currently have long-term disabilities following TBI, resulting in an estimated $60 billion in health care expenditures annually.[1,11] For individuals, the significant financial burden that results from TBI depends on factors including duration of acute care, duration of rehabilitation and long-term care, therapies, prescription costs, loss of employment, and need for medical equipment. The U.S. National Institutes of Health estimates that lifetime care for a survivor with severe TBI can cost between U.S.$ 600 000 and U.S.$ 1 875 000.[12]

In light of the growing TBI epidemic, use of biomarkers for the diagnosis of TBI and its potential sequelae is essential to optimize patient care and help improve clinical outcome. Additionally, biomarkers could be potentially useful to identify individuals at risk for secondary injury following the initial impact. Such individuals are more prone to subsequent brain damage and long-term disability if they are released prematurely to return to high risk behaviors and activities. Biomarkers could thus play an important role in the care of patients with TBI.

1.3 Diagnosis

Currently, a diagnosis of TBI is based on medical history, findings on neurological examination, clinical assessment scales, and neuroimaging, such as computed tomography (CT) of the head and magnetic resonance imaging (MRI), single-photon emission computed tomography (SPECT), or positron emission tomography (PET) of the brain. While a diagnosis of moderate to severe TBI is often self evident from patient history and associated signs of injury or abnormalities on neurological examination and neuroimaging, a diagnosis of mild TBI frequently remains difficult. Multiple classification schemes for mild TBI exist; although there is some overlap, there is also controversy regarding the qualifying diagnostic criteria, such as presence of loss of consciousness and duration of changes in sensorium. It has been generally accepted that mild TBI is defined by an injury that is significantly severe enough to cause transient confusion, impairment of memory function, and disorientation with or without accompanying loss of consciousness.[13] Following these transient initial findings, the patient may first seem to have returned to his/her normal neurological baseline and has a normal neurological examination. However, frequently patients with mild TBI then develop non-specific symptoms such as fatigue, headaches, visual disturbances, memory difficulties, inability to concentrate, inattentiveness, sleep disturbances, dizziness and imbalance, emotional disturbances, depression and/or seizures, which can occur immediately following the injury or after a delay of days to several weeks. These symptoms may be subtle and can go unrecognized until the patient or others notice work performance difficulties, personality changes, or inability to perform complex tasks. Family or patients may notice increased sensitivity to light, irritability, mood changes, confusion, and mental slowing. However, as the neurological examination and neuroimaging studies are frequently normal, the diagnosis of mild TBI remains challenging and there are no objective prognostic predictors following mild TBI.

Multiple clinical assessment scales to grade the severity of TBI have been utilized, including the Cantu or the American Academy of Neurology grading systems.[14–16] The most widely known and replicated scale is the Glasgow Coma Scale (GCS), which assesses best motor, eye opening, and verbal response that can be observed spontaneously and following verbal or painful stimulation. While the GCS has proven its utility in the clinical management and prognosis of severe TBI patients, it cannot provide information about the pathophysiological mechanisms responsible for a patient's neurological deficits. In addition, specific patient populations are difficult to assess with the GCS, particularly those who suffer from mild or moderate TBI. Many mild head trauma patients with a GCS between 13 and 15 are also intoxicated with drugs and alcohol that frequently confound clinical neuropsychological examinations.[17] Head injuries may also be overlooked in multi-trauma patients.[18]

Neuroimaging techniques can provide additional clinical information, but the diagnostic capabilities of these techniques are limited by their sensitivities, accessibility, high capital costs, and service requirements. CT is the most rapid and widely available neuroimaging technique, yet its usefulness in detecting diffuse brain damage from axonal shear injury is limited. In critically ill patients and in those with contraindications to undergo MRI, brain MRI frequently cannot be obtained. Furthermore, limited availability, high cost, and the time to acquire and to analyze images limit the use of functional MRI, PET, and SPECT scans. SPECT and PET are capable of detecting regional changes in blood flow but cannot necessarily detect structural damage. Furthermore, MRI and CT often do not provide information that can predict outcome of mild TBI and concussion (for details on neuroimaging in TBI, see also Chapter 2). Typical findings on brain MRI following moderate TBI are shown in Figure 1.1.

A rare, but potentially life-threatening, complication following TBI that is primarily observed in athletes is the second impact syndrome. Individuals

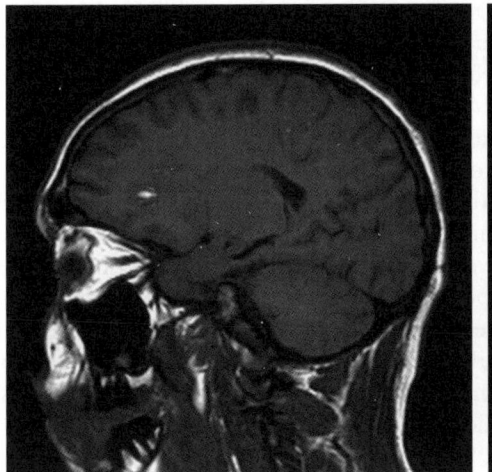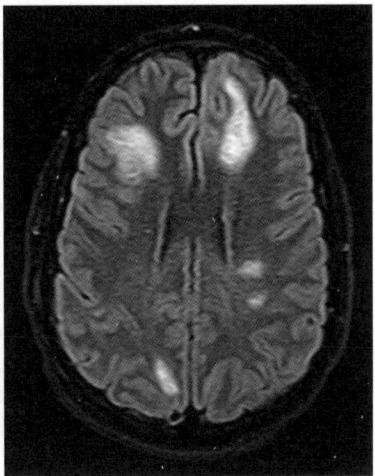

Figure 1.1 Brain MRI showing diffuse axonal shear injury following moderate to severe TBI (left panel: sagittal T1 weighted MRI; right panel: fluid attenuated inversion recovery imaging). There are multiple hemorrhagic foci at the gray/white matter junction in both cerebral hemispheres with nonhemorrhagic contusions in the splenium of the corpus callosum consistent with diffuse axonal brain injury.

who have sustained an acute TBI frequently develop a delayed disruption of the cerebrovascular autoregulation leading to diffuse cerebral hyperemia, a hypermetabolic state, and dysfunction of the blood-brain barrier. During this phase, the brain is highly susceptible to severe tissue damage from recurrent head injury. A second impact during this critical phase can cause malignant cerebral edema, which can be refractory to any treatment and may cause brain herniation, devastating secondary cerebral injury, and death. It would be of great clinical importance to identify individuals at risk for second impact syndrome. Biomarkers could potentially be useful to identify these individuals and help guide the decision of when it is safe to return the athlete to play or for a soldier to return to active duty.

Long-term sequelae following TBI include vision changes, speech and cognitive impairment, personality and mood changes, abnormal behavior, paralysis, sensory deficits, gait dysfunction, seizures, chronic pain, and various degrees of disability. Biomarkers may be useful in predicting these sequelae and to help optimize treatment and preventive strategies following TBI.

1.4 Potential Clinical Applications of TBI Biomarkers

Detection of markers of brain tissue injury is complicated by the presence of the blood-brain barrier (BBB). Any potentially useful biomarker has to penetrate this barrier to become released into the bloodstream, cerebrospinal fluid, or other body fluid in sufficient quantities to be reliably identified. From a practical standpoint, a simple point of care blood test would be ideal. Although the concept is intuitively simple, markers of brain injury are difficult to detect because of the complexity of the pathophysiological changes that take place in the brain following TBI, including the activation of multiple molecular pathways of neurodegeneration and neuroregeneration, metabolic compromise, oxidative stress, inflammation, vascular dysfunction, and secondary tissue injury, making it difficult to isolate brain-specific biomarkers.

An ideal TBI biomarker must be brain-tissue specific, sensitive, and should be detectable in blood within minutes of the onset of symptoms; ideally, it should also be inexpensive and easy to measure. Moreover, the concentration of the biomarker should correspond to lesion size, location, and functional outcome. Its predictive value needs to be compared to the sensitivity and specificity of brain MRI, clinical assessment scales, and the neurocognitive function tests that serve as reference standards for the diagnosis and prognosis of TBI. Currently, there is no sufficiently specific and sensitive single biomarker or panel of biomarkers validated by large clinical trials. Thus, many questions regarding the clinical utility of biomarkers for TBI remain.

Reliable and cost-effective biomarkers could help establish a diagnosis of TBI, especially in those with mild cases, where TBI which often goes undetected. In patients who have no immediate access to brain MRI or have contraindications to receiving an MRI or advanced neuroimaging, biomarkers could improve diagnostic certainty, allowing timely identification and triaging for further management at specialized medical centers. Furthermore, in the acute setting and during early clinical follow-up, they could be useful predictors of clinical outcome and potential complications, such as the development of malignant cerebral edema or ischemia that can follow TBI. They could also serve useful in identifying individuals who are at risk for second impact syndrome, guiding the decision of when someone who has suffered a TBI can safely assume his/her activity.

If biomarkers are proven to be sensitive and specific for diagnosis of mild TBI, diagnostic errors could be reduced, adverse outcomes could be minimized, and patients requiring hospital admission could be clearly identified and treated. All of these measures could theoretically help reduce health care expenditures and improve the quality of patient management. However, it remains to be shown whether TBI biomarkers have sufficient predictive value and/or produce likelihood ratios that are acceptable for standard use in clinical practice.

1.5   Human Biomarkers for TBI

Types of biomarkers for TBI currently include – but are not limited to – proteins, enzymes, protein degradation products, cytokines and, more recently, microRNAs. For an overview of human studies of CSF and blood biomarkers for TBI, see Table 1.2. This table serves as an overview of the most commonly studied biomarkers in humans and is not meant to be complete; the interested reader is encouraged to refer to the existing literature for more in-depth information.

Great variability in methodologies and study populations exist in the current TBI biomarker literature. Different outcome measures, sampling times,

Table 1.2 Summary of results obtained from clinical TBI biomarker studies. The table does not provide a complete listing of all TBI biomarkers, but is intended to list some of the most commonly studied.

Type	Biomarker	Source	Mechanism/Target	Severity of TBI	Study Models Used	Major Findings
Proteins and Enzymes	S100β	CSF and serum	Glial and blood brain barrier dysfunction marker	Moderate and severe	Prospective Observational Restrospective Database	~ Elevated CSF and serum levels in severe TBI ~ Not distinguishable from multi-trauma ~ Serum levels related to outcome ~ Sensitive but not specific
		Serum		Mild	Prospective Observational	~ Some studies showed elevated in serum ~ Sensitive but not specific ~ Conflicting data
	Neuron-specific enolase (NSE)	CSF and serum	Glycolytic enzyme, marker of axonal injury	Moderate and severe	Prospective Observational Retrospective Database	~ Elevated serum and CSF levels in severe TBI ~ Perhaps an indicator of neuroinflammation
				Mild	Prospective Observational	~ Serum levels influenced by hemolysis ~ Conflicting data for mild TBI
	Myelin basic protein (MBP)	Serum and CSF	Demyelination marker	Severe	Prospective Observational	~ Elevated serum levels in severe TBI ~ In severe TBI CSF levels correlated with outcome ~ Poor specificity
		Serum		Mild	Prospective Observational	~ In one study serum levels elevated after mild TBI
	Glial fibrillary acidic protein (GFAP)	Serum	Gliosis marker	Severe	Prospective Observational Retrospective Database	~ Elevated serum levels in severe TBI ~ Serum levels correlated with mortality and outcome ~ Distinguishable from multitrauma
				Mild	Prospective Observational	~ Preliminary data suggests elevated serum levels in mild TBI

Table 1.2 (Continued)

Type	Biomarker	Source	Mechanism/ Target	Severity of TBI	Study Models Used	Major Findings
	Ubiquitin C-terminal hydrolase (UCH-L1)	CSF and serum	Neural cell body marker	Severe	Prospective Observational	~ In severe TBI CSF levels correlated with GCS ~ Serum levels correlated with TBI and blood brain barrier dysfunction
		Serum		Mild	Prospective Observational	~ Preliminary data suggests elevated serum levels in mild TBI
	Neurofilament-L (NFL) and neuro-filament-H (NF-H)	CSF and serum	Axonal injury marker	Severe	Prospective Observational	~ NFL level elevated in CSF in severe TBI ~ NFH level elevated in CSF and serum in severe TBI ~ NFH correlated with outcome ~ NFH and NFL not correlated with blood brain barrier dysfunction
		CSF		Mild	Prospective Observational	~ NFL not elevated following mild TBI
	Brain Fatty Acid Binding Protein (B-FABP)	Serum	Glial cell break-down marker	Mild	Prospective Observational	~ Elevated serum levels in mild TBI

Degradation Products					
αII-spectrin break-down products (SBDP150, SBDP145, SBDP120)	CSF and Serum	Neural necrosis/apoptosis, axonal injury	Severe	Prospective Observational Retrospective Database	~ Elevated CSF levels in severe TBI ~ CSF levels correlated with outcome ~ Elevated serum levels in severe TBI
Cleaved-tau (C-tau)	CSF and serum	Axonal injury marker	Severe Mild	Prospective Observational Prospective Observational	~ Elevated CSF levels in severe TBI ~ In CSF correlated with outcome and mortality ~ In CSF distinguishable from neurological disease and trauma ~ No correlation with serum levels in severe TBI ~ Poor sensitivity and specificity for mild TBI
Amyloid-β_{1-42}	CSF	Neural necrosis/apoptosis, axonal injury	Severe	Prospective Observational	~ Elevated CSF levels in severe TBI in some studies ~ Conflicting data

inclusion/exclusion criteria, types of "healthy controls," definitions of severity of TBI, diagnostic biomarker cut-off levels, and measurement methods for biomarkers have been used, making it difficult to compare the efficacy and reliability of multiple biomarkers described in the different human studies. In addition, there is very limited research on mild TBI to date. Thus far, the large majority of research in TBI biomarkers has focused on S100beta, neuron-specific enolase (NSE), myelin basic protein (MBP), and glial fibrillary acidic protein (GFAP).

One of the most widely researched biomarkers for TBI is S100beta, a calcium-binding protein that can be detected in serum and cerebrospinal fluid (CSF) following TBI. Within the human body, S100beta has been found in the cytoplasm of astroglia; in Schwann cells, adipocytes, and chondrocytes; and in melanoma cells; therefore, it cannot be considered a brain-specific biomarker. It is believed that S100beta is released into the central nervous system (CNS) following glial cell damage. If the brain injury is accompanied by breakdown of the BBB, sufficient quantities of S100beta will be released into the bloodstream, allowing detection. Following severe TBI, S100beta serum levels are significantly elevated and correlate well with poor clinical outcome and abnormal head CT findings.[19,20] In most studies, the biomarker has generally been found to peak within the first 12 hours following injury, although this time course has been variable between studies.[19] Overall, S100beta serum levels have high sensitivity but only poor specificity for diagnosis of moderate to severe TBI.[21–23] In contrast to many other potential biomarkers of TBI, studies have been conducted to assess its diagnostic use for mild TBI. Some studies have shown a correlation between elevated S100beta in serum following mild TBI and clinical outcome, but other studies could not confirm these findings and revealed conflicting results.[24,25] Cut-off levels for diagnosis of mild TBI also varied widely between studies.[25] The most significant drawback to the diagnostic use of S100beta is that it is not specific for TBI and cannot distinguish between brain injury and peripheral multi-trauma.[19,20,24–26] Elevated serum S100beta levels are not only observed after TBI, but may also reflect release due to neurodegenerative processes associated with aging or breakdown of the BBB from other etiologies.[19,27] Ultimately, S100beta lacks the specificity for TBI to make it a reliable diagnostic tool for clinical use.

NSE (neuron-specific enolase) has been studied in mild, moderate, and severe TBI, and can be measured in blood and CSF. NSE is a highly stable glycolytic enzyme that is primarily found in the cytoplasm of neurons. It is a marker of axonal injury and has been shown to be correlated with severe TBI, but seems to be inadequate for detecting mild and moderate TBI.[28] Low levels of NSE have been shown to be present in CSF and serum of healthy individuals.[29,30] In severe TBI, NSE peaks in serum within 12 hours after injury.[25] Studies in severe TBI have shown conflicting results in relating NSE CSF and serum levels to clinical outcome, but most studies indicate that NSE is not correlated with measures of long-term clinical outcome.[31] Following mild TBI, elevated serum NSE levels showed relatively high diagnostic specificity but low sensitivity.[32] NSE is eliminated from the body slowly, complicating its use as a marker for secondary versus primary injury.[33] While NSE is primarily found in neurons, it can also be released into the bloodstream by certain tumors, with hemolysis, and with normal aging; therefore, it is not clinically useful when used as a sole diagnostic test for TBI.[27,31,34,35] Serum levels of NSE have shown some promise when used in conjunction with S100beta, whereby elevation of both proteins have been correlated with all severities of TBI in children; however, when both biomarkers were combined, the sensitivity was 80% and specificity, 73%.[36]

MBP (myelin basic protein) is specific to myelin and is released into serum following neurological injury and during different neurological disease processes. Most studies have shown elevated levels of MBP following moderate and severe TBI, but only one study has shown increased levels following mild TBI.[37–39] The release of MBP into serum depends on BBB damage, so the time course of MBP in serum is influenced by severity and extent of this damage.[40] In several studies of severe TBI, serum MBP levels were correlated with mortality.[31] However, the utility of MBP is ultimately limited by its poor specificity. MBP is a common marker of demyelination, which is also associated with Alzheimer's disease, multiple sclerosis, and other disease processes, and thus MBP is not specific for diagnosis of TBI. In addition to preexisting conditions causing elevated baseline levels, the impact of aging and technical difficulties in measuring MBP in serum make MBP an unreliable diagnostic tool for clinical practice.[26,27,31,41–43]

More promising research has been shown for GFAP (glial fibrillary acidic protein), which is found only in CNS glial cells. GFAP is a structural protein of astroglia and can be found in CSF and serum following TBI.[44] In serum, GFAP is mainly derived from the CNS, and usually peaks within 1 to 4 days following severe TBI.[25,26] In moderate and severe TBI, GFAP measured in serum and CSF has been shown to have the potential to indicate clinical outcome, and to predict morbidity and mortality.[33,44–49] However, the release of GFAP may depend on BBB damage.[46] GFAP serum levels are best able to predict mortality at earlier – rather than later – time points after severe TBI. GFAP appears to be brain specific, as it is not elevated following multi-trauma.[44,47] Preliminary data from the "BANDITS" trial suggest that serum GFAP may be significantly correlated with the diagnosis of mild TBI; however, further studies will be necessary to confirm these early findings.[50]

There has been growing interest in the use of ubiquitin C-terminal hydrolase-L1 (UCH-L1) as a biomarker of TBI. Emerging studies show that UCH-L1 is a neuron-specific enzyme, which may be useful as a marker of neural cell body injury. It can be found in CSF and serum following TBI, and CSF levels were significantly correlated with severity of brain injury.[33,51] However, early analysis in a small patient population indicated that the time course of UCH-L1 following TBI is still not well characterized.[52] Additionally, serum UCH-L1 has been correlated with BBB dysfunction following moderate to severe TBI. With a half-life estimated to be 7 to 9 hours in both CSF and serum, UCH-L1 may be a promising early marker of TBI, peaking in both serum and CSF within 48 hours after injury.[53] Preliminary results from the "BANDITS" trial suggest that UCH-L1 levels may be significantly elevated in serum following mild TBI.[50]

Other potential clinically useful markers of TBI include alpha II-spectrin breakdown products (SBDPs), cleaved-tau (C-tau), the light and heavy chain neurofilaments (NFL and NFH, respectively), brain fatty acid binding protein (B-FABP), and amyloid-beta$_{1-42}$. The degradation of alpha II-spectrin as a result of axonal injury produces three main products in CSF: SBDP150, SBDP145, and SBDP120, which are markers of necrosis, mediated by calpain, and apoptosis, mediated by caspases. All three have been found to peak in CSF within 6 hours to 3 days after TBI.[40,54] CSF levels of SBDP150 and SBDP145 have been found to correlate with severe TBI and clinical outcome.[33,40,52] Elevated serum levels of SBDP150 have also been found following severe TBI.[52] To our knowledge, no data on SBDPs in mild TBI in adults have yet been published.

C-tau results from the cleavage of microtubule-associated protein tau, which is a phosphoprotein responsible for assembling microtubule bundles of the axonal cytoskeleton. In severe TBI, C-tau has been significantly associated with TBI within the first 24 hours following injury, predicting clinical outcome with relatively high sensitivity and specificity. In mild TBI, the data for C-tau have been much less promising. Few studies have revealed any correlation between C-tau and mild TBI and, when correlations were found, they had poor sensitivity and specificity.[25]

NFL and NFH (light and heavy chain neurofilaments) are part of the axonal cytoskeleton, where they are involved in the regulation of axonal organization and function, and are broken down by caspase-3 and calpain following axonal injury. NFL is released into CSF first and, although human data is limited to a single study with only three severe and three mild TBI patients, it has been shown to be stable and significantly elevated following severe TBI, but not following mild TBI, acute ischemic stroke, transient ischemic attack, or subarachnoid hemorrhage.[55] To our knowledge, only one human study in TBI has measured NFH in CSF, and was not found to be significantly correlated with severe TBI. However, these results need to be interpreted with caution, as the study was conducted in a small population.[53]

B-FABP (brain fatty acid binding protein) is found only in brain tissue and is released into the CSF following axonal damage; it is released into serum only after BBB damage. Currently, only one human trial has studied B-FABP following TBI, which found a correlation between B-FABP levels in serum and diagnosis of mild TBI.[56,57]

Amyloid-beta$_{1-42}$ is a breakdown product of amyloid precursor protein, a cellular adhesion protein. Amyloid-beta$_{1-42}$ has long been considered a marker of Alzheimer's disease, but it has also been shown to be significantly elevated following TBI. In some studies, amyloid-beta$_{1-42}$ levels in CSF following severe TBI have been shown to be significantly elevated compared to healthy controls and subjects with Alzheimers disease. In contrast, other studies found that amyloid-beta$_{1-42}$ levels are significantly decreased following TBI.[25,58–65] Interestingly, plasma levels of amyloid-beta$_{1-42}$ do not seem to reflect CSF levels of amyloid-beta$_{1-42}$ or TBI severity.[62] Obviously, further research is needed to understand better the release and use of SBDPs, C-tau, NFL, NFH, B-FABP, and amyloid-beta$_{1-42}$ following TBI.

Current research interest focuses on non-specific markers of TBI, including cytokines and other markers of inflammation and apoptosis, such as cytochrome C, Bcl-2 proteins, and YKL-40, which are limited in application.[66–69] Such markers may be helpful in the prediction of secondary injury, but cut-offs have not been established. While these biomarkers are not individually clinically diagnostic of TBI, they may have clinical utility when combined into biomarker panels and if they are used in conjunction with other diagnostic tests for TBI. It should be noted that while inflammatory markers may be helpful, they are easily compromised by trauma in other areas of the body, infection, and other inflammatory processes, making them less useful for moderate to severe TBI as they are not specific to cerebral injury.

1.6 Future Directions

New biomarkers of TBI and advanced MRI-based neuroimaging technologies are currently being evaluated. Advances in MRI and other imaging technologies have led to greater understanding of the impact of TBI, especially mild TBI, with the expectations of improved diagnostic certainty in the identification of mild TBI and in the prediction of clinical outcome.[70–72] The pathophysiology of mild TBI may soon be better understood using newer MRI modalities such as MR spectroscopy, which allows researchers to monitor metabolic changes that occur after TBI and during recovery of brain function. Diffusion tensor MRI also may improve insight into structural damage following TBI, such as axonal shear injury, and into processes of brain plasticity that follow the injury (for a detailed discussion, see Chapter 2).

Assessment and validation of several new proteins and biomarker panels for the diagnosis of TBI, as well as the rapidly advancing field of genomics, may introduce new diagnostic tools for TBI. Fragments of glutamate receptors, which are widely distributed throughout the CNS, are being investigated for diagnosis of acute TBI and its prognosis to determine their utility to serve as brain-specific markers of injury and TBI independent of injury severity.[73,74]

Recently, proteins 14-3-3beta and 14-3-3zeta have shown promise as useful markers of acute TBI. Protein 14-3-3 in CSF is already being used for diagnosis of Creutzfeldt-Jacob disease.[75] Most of the research with these proteins in TBI has been conducted only in animal models and will require validation in human TBI populations. The 14-3-3 proteins have also been used in combination with other proteins suspected to be markers of TBI. However, the use of these biomarker panels to diagnose TBI has revealed shortcomings in sensitivity and specificity. Furthermore, it is currently not known how well patterns of elevated protein panels correlate with lesion size, lesion location, and outcome.[76–78] Further studies are needed to determine the time course of these biomarker panels and to correlate these findings with clinical, neuroimaging, and pathological changes.

The field of TBI biomarkers is advancing rapidly due to recent discoveries in proteomics and genomics. New proteomic techniques can help identify multiple potential TBI biomarker candidates. Genomic studies target millions of genes and their protein products in a short period of time, allowing the study of patterns in gene up-regulation and down-regulation involved in response to brain injury and cell repair mechanisms. For example, circulating plasma microRNA levels are altered following TBI and could be useful as a marker of TBI.[79] Genomic expression patterns for pathways of inflammation and neurodegeneration, such as cytokines, chemokines, and growth factors, have been studied in animal models of TBI. Meanwhile, limited research has been done to date on genetic expression following TBI in humans.[80] Most of these early studies have been done in brain tissue of single TBI patients.[81–83] Study of gene expression patterns in peripheral blood following TBI is emerging; currently, specificity and sensitivity of these patterns are not sufficient for routine clinical practice.[79] Peripheral blood genomic profiling will also allow improved understanding of the complex mechanisms involved in the pathophysiology and repair processes following TBI. For example, they might be especially helpful in characterizing mild TBI, in which the pathophysiology is still poorly understood. In the future, development in stem cell research may allow the detection of markers of factors involved in the promotion of neuronal stem cell-progenitor cells and the differentiation following activation of microglia after TBI.[84]

In summary, the field of biomarkers for TBI is rapidly evolving and while several biomarkers and new genomic approaches show promise in early studies, large clinical validation studies will be required to assess their utility in clinical practice.

References

1. C. L. Armstrong, L. Morrow and T. Morris, in *Handbook of Medical Neuropsychology*, Springer, New York, 2010, p. 17.
2. A. Bhardwaj, M. A. Mirski, G. S. F. Ling and S. A. Marshall, in *Handbook of Neurocritical Care*, Springer, New York, 2010, p. 307.
3. J. A. Norton, P. S. Barie, R. R. Bollinger, A. E. Chang, S. F. Lowry, S. J. Mulvihill, H. I. Pass, R. W. Thompson, K. Chappie and R. Hartl, in *Surgery*, Springer, New York, 2008, p. 461.
4. C. Werner and K. Engelhard, *Br. J. Anaesth.*, 2007, **99**, 4.
5. X. L. Faul, M. M. Wald and V. Coronado, *Traumatic Brain Injury in the United States: Emergency Department Visits, Hospitalizations, and Deaths, 2002–2006*, CDC, National Center for Injury Prevention and Control, Atlanta, GA, 2010.
6. V. G. Coronado, L. Xu, S. V. Basavaraju, L. C. McGuire, M. M. Wald, M. D. Faul, B. R. Guzman and J. D. Hemphill, *M.M.W.R. Surveill. Summ.*, 2011, **60**, 1.
7. http://www.ninds.nih.gov/disorders/tbi/detail_tbi.htm (last accessed September 2011).
8. D. Warden, *J. Head Trauma Rehabil.*, 2006, **21**, 398.
9. http://www.dvbic.org/TBI-Numbers.aspx
10. A. I. Maas, N. Stocchetti and R. Bullock, *Lancet Neurol.*, 2008, **7**, 728.

11. E. Finkelstein, P. S. Corso and T. R. Miller, *The Incidence and Economic Burden of Injuries in the United States*, Oxford University Press, Oxford, 2006.
12. Rehabilitation of Persons with Traumatic Brain Injury. NIH Consensus Statement Online 1998 Oct 26–28, **16**(1), 1 (last accessed September 2011).
13. R. T. Seel, M. Sherer, J. Whyte, D. I. Katz, J. T. Giacino, A. M. Rosenbaum, F. M. Hammond, K. Kalmar, T. L. Pape, R. Zafonte, R. C. Biester, D. Kaelin, J. Kean and N. Zasler, *Arch. Phys. Med. Rehabil.*, 2010, **91**, 1795.
14. http://www.cdc.gov/nchs/data/icd9/Sep08TBI.pdf (last accessed September 2011).
15. R. Cantu, *Physician Sportsmed.*, 1986, **14**, 75.
16. J. P. Kelly and J. H. Rosenberg, *Neurology*, 1997, **48**, 575.
17. J. D. Golan, J. Marcoux, E. Golan, R. Schapiro, K. M. Johnston, M. Maleki, S. Khetarpal and L. Jacques, *J. Trauma*, 2007, **63**, 365.
18. M. S. Greenberg, *Handbook of Neurosurgery*, Thieme Medical Publishers, New York, 2010.
19. A. Kleindienst and M. R. Bullock, *J. Neurotrauma*, 2006, **23**, 1185.
20. B. J. Zink, J. Szmydynger-Chodobska and A. Chodobski, *Psychiatr. Clin. North Am.*, 2010, **33**, 741.
21. A. Raabe, C. Grolms and V. Seifert, *Br. J. Neurosurg.*, 1999, **13**, 56.
22. R. D. Rothoerl, C. Woertgen, M. Holzschuh, C. Metz and A. Brawanski, *J. Trauma*, 1998, **45**, 765.
23. C. Woertgen, R. D. Rothoerl, C. Metz and A. Brawanski, *J. Trauma*, 1999, **47**, 1126.
24. S. M. Bloomfield, J. McKinney, L. Smith and J. Brisman, *Neurocrit. Care*, 2007, **6**, 121.
25. E. Kovesdi, J. Luckl, P. Bukovics, O. Farkas, J. Pal, E. Czeiter, D. Szellar, T. Doczi, S. Komoly and A. Buki, *Acta Neurochir. (Wien)*, 2010, **152**, 1.
26. K. J. Lamers, P. Vos, M. M. Verbeek, F. Rosmalen, W. J. van Geel and B. G. van Engelen, *Brain Res. Bull.*, 2003, **61**, 261.
27. B. G. van Engelen, K. J. Lamers, F. J. Gabreels, R. A. Wevers, W. J. van Geel and G. F. Borm, *Clin. Chem.*, 1992, **38**, 813.
28. I. M. Skogseid, H. K. Nordby, P. Urdal, E. Paus and F. Lilleaas, *Acta Neurochir. (Wien)*, 1992, **115**, 106.
29. U. E. Pleines, M. C. Morganti-Kossmann, M. Rancan, H. Joller, O. Trentz and T. Kossmann, *J. Neurotrauma*, 2001, **18**, 491.
30. P. J. Marangos and D. E. Schmechel, *Annu. Rev. Neurosci.*, 1987, **10**, 269.
31. T. Ingebrigtsen and B. Romner, *J. Trauma*, 2002, **52**, 798.
32. T. Ingebrigtsen and B. Romner, *Restor. Neurol. Neurosci.*, 2003, **21**, 171.
33. S. Mondello, U. Muller, A. Jeromin, J. Streeter, R. L. Hayes and K. K. Wang, *Expert Rev. Mol. Diagn.*, 2011, **11**, 65.
34. H. Jonsson, P. Johnsson, M. Backstrom, C. Alling, C. Dautovic-Bergh and S. Blomquist, *BMC Neurol.*, 2004, **4**, 24.
35. L. E. Pelinka, H. Hertz, W. Mauritz, N. Harada, M. Jafarmadar, M. Albrecht, H. Redl and S. Bahrami, *Shock*, 2005, **24**, 119.
36. R. P. Berger, P. D. Adelson, M. C. Pierce, T. Dulani, L. D. Cassidy and P. M. Kochanek, *J. Neurosurg.*, 2005, **103**, 61.

37. Y. Yamazaki, K. Yada, S. Morii, T. Kitahara and T. Ohwada, *Surg. Neurol.*, 1995, **43**, 267; discussion 270.
38. R. P. Berger, S. R. Beers, R. Richichi, D. Wiesman and P. D. Adelson, *J. Neurotrauma*, 2007, **24**, 1793.
39. T. W. Noseworthy, B. J. Anderson, A. F. Noseworthy, A. Shustack, R. G. Johnston, K. C. Petruk and T. A. McPherson, *Crit. Care Med.*, 1985, **13**, 743.
40. J. Li, X. Y. Li, D. F. Feng and D. C. Pan, *J. Trauma*, 2010, **69**, 1610.
41. J. Matias-Guiu, J. Martinez-Vazquez, A. Ruibal, R. Colomer, M. Boada and A. Codina, *Acta Neurol. Scand.*, 1986, **73**, 461.
42. J. Matias-Guiu, J. M. Martinez-Vazquez, A. Ruibal and A. Codina, *Acta Neurol. Scand.*, 1986, **73**, 203.
43. J. Matias-Guiu, A. Ruibal, J. M. Martinez-Vazquez, R. Colomer and A. Codina, *Clin. Chem.*, 1986, **32**, 915.
44. K. Nylen, M. Ost, L. Z. Csajbok, I. Nilsson, K. Blennow, B. Nellgard and L. Rosengren, *J. Neurol. Sci.*, 2006, **240**, 85.
45. U. Missler, M. Wiesmann, G. Wittmann, O. Magerkurth and H. Hagenstrom, *Clin. Chem.*, 1999, **45**, 138.
46. P. E. Vos, B. Jacobs, T. M. Andriessen, K. J. Lamers, G. F. Borm, T. Beems, M. Edwards, C. F. Rosmalen and J. L. Vissers, *Neurology*, 2010, **75**, 1786.
47. L. E. Pelinka, A. Kroepfl, R. Schmidhammer, M. Krenn, W. Buchinger, H. Redl and A. Raabe, *J. Trauma*, 2004, **57**, 1006.
48. W. J. van Geel, H. P. de Reus, H. Nijzing, M. M. Verbeek, P. E. Vos and K. J. Lamers, *Clin. Chim. Acta*, 2002, **326**, 151.
49. J. Hanrieder, M. Wetterhall, P. Enblad, L. Hillered and J. Bergquist, *J. Neurosci. Methods*, 2009, **177**, 469.
50. J. Streeter, R. L. Hayes and K. K. W. Wang, *Sensing Technologies for Global Health, Military Medicine, Disaster Response, and Environmental Monitoring; and Biometric Technology for Human Identification VIII*, Eds. S. O. Southern, K. N. Montgomery, C. W. Taylor, B. H. Weigl, B. V. K. V. Kumar, S. Prabhakar and A. A. Ross, Proceedings of SPIE, 2011, 8029, 80290N. doi: 10.1117/12.885615.
51. L. Papa, L. Akinyi, M. C. Liu, J. A. Pineda, J. J. Tepas, 3rd, M. W. Oli, W. Zheng, G. Robinson, S. A. Robicsek, A. Gabrielli, S. C. Heaton, H. J. Hannay, J. A. Demery, G. M. Brophy, J. Layon, C. S. Robertson, R. L. Hayes and K. K. Wang, *Crit. Care Med.*, 2010, **38**, 138.
52. R. Siman, N. Toraskar, A. Dang, E. McNeil, M. McGarvey, J. Plaum, E. Maloney and M. S. Grady, *J. Neurotrauma*, 2009, **26**, 1867.
53. B. J. Blyth, A. Farahvar, H. He, A. Nayak, C. Yang, G. Shaw and J. J. Bazarian, *J. Neurotrauma*, 2011; Epub 2011 Aug 08.
54. J. A. Pineda, S. B. Lewis, A. B. Valadka, L. Papa, H. J. Hannay, S. C. Heaton, J. A. Demery, M. C. Liu, J. M. Aikman, V. Akle, G. M. Brophy, J. J. Tepas, K. K. Wang, C. S. Robertson and R. L. Hayes, *J. Neurotrauma*, 2007, **24**, 354.
55. W. J. Van Geel, L. E. Rosengren and M. M. Verbeek, *J. Immunol. Methods*, 2005, **296**, 179.

56. M. M. Pelsers and J. F. Glatz, *Clin. Chem. Lab. Med.*, 2005, **43**, 802.
57. M. M. Pelsers, T. Hanhoff, D. Van der Voort, B. Arts, M. Peters, R. Ponds, A. Honig, W. Rudzinski, F. Spener, J. R. de Kruijk, A. Twijnstra, W. T. Hermens, P. P. Menheere and J. F. Glatz, *Clin. Chem.*, 2004, **50**, 1568.
58. A. D. Kay, A. Petzold, M. Kerr, G. Keir, E. Thompson and J. A. Nicoll, *J. Neurotrauma*, 2003, **20**, 943.
59. N. Marklund, K. Blennow, H. Zetterberg, E. Ronne-Engstrom, P. Enblad and L. Hillered, *J. Neurosurg.*, 2009, **110**, 1227.
60. M. R. Emmerling, M. C. Morganti-Kossmann, T. Kossmann, P. F. Stahel, M. D. Watson, L. M. Evans, P. D. Mehta, K. Spiegel, Y. M. Kuo, A. E. Roher and C. A. Raby, *Ann. N.Y. Acad. Sci.*, 2000, **903**, 118.
61. C. A. Raby, M. C. Morganti-Kossmann, T. Kossmann, P. F. Stahel, M. D. Watson, L. M. Evans, P. D. Mehta, K. Spiegel, Y. M. Kuo, A. E. Roher and M. R. Emmerling, *J. Neurochem.*, 1998, **71**, 2505.
62. A. Olsson, L. Csajbok, M. Ost, K. Hoglund, K. Nylen, L. Rosengren, B. Nellgard and K. Blennow, *J. Neurol.*, 2004, **251**, 870.
63. G. Franz, R. Beer, A. Kampfl, K. Engelhardt, E. Schmutzhard, H. Ulmer and F. Deisenhammer, *Neurology*, 2003, **60**, 1457.
64. S. Magnoni and D. L. Brody, *Arch. Neurol.*, 2010, **67**, 1068.
65. D. L. Brody, S. Magnoni, K. E. Schwetye, M. L. Spinner, T. J. Esparza, N. Stocchetti, G. J. Zipfel and D. M. Holtzman, *Science*, 2008, **321**, 1221.
66. A. K. Wagner, K. B. Amin, C. Niyonkuru, B. A. Postal, E. H. McCullough, H. Ozawa, C. E. Dixon, H. Bayir, R. S. Clark, P. M. Kochanek and A. Fabio, *J. Cereb. Blood Flow Metab.*, 2011, **31**, 1886.
67. M. A. Satchell, Y. Lai, P. M. Kochanek, S. R. Wisniewski, E. L. Fink, N. A. Siedberg, R. P. Berger, S. T. DeKosky, P. D. Adelson and R. S. Clark, *J. Cereb. Blood Flow Metab.*, 2005, **25**, 919.
68. D. Bonneh-Barkay, P. Zagadailov, H. Zou, C. Niyonkuru, M. Figley, A. Starkey, G. Wang, S. J. Bissel, C. A. Wiley and A. K. Wagner, *J. Neurotrauma*, 2010, **27**, 1215.
69. D. M. Stein, A. Lindell, K. R. Murdock, J. A. Kufera, J. Menaker, K. Keledjian, G. V. Bochicchio, B. Aarabi and T. M. Scalea, *J. Trauma*, 2011, **70**, 1096.
70. T. A. Huisman, L. H. Schwamm, P. W. Schaefer, W. J. Koroshetz, N. Shetty-Alva, Y. Ozsunar, O. Wu and A. G. Sorensen, *AJNR Am. J. Neuroradiol.*, 2004, **25**, 370.
71. S. Signoretti, R. Vagnozzi, B. Tavazzi and G. Lazzarino, *Neurosurg. Focus*, 2010, **29**, E1.
72. S. Marino, E. Zei, M. Battaglini, C. Vittori, A. Buscalferri, P. Bramanti, A. Federico and N. De Stefano, *J. Neurol. Neurosurg. Psychiatry*, 2007, **78**, 501.
73. E. A. Kharlamov, E. Lepsveridze, M. Meparishvili, R. O. Solomonia, B. Lu, E. R. Miller, K. M. Kelly and Z. Mtchedlishvili, *Epilepsy Res.*, 2011, **95**, 20.
74. P. Luo, F. Fei, L. Zhang, Y. Qu and Z. Fei, *Brain Res. Bull.*, 2011, **85**, 313.

75. A. W. Lemstra, M. T. van Meegen, J. P. Vreyling, P. H. Meijerink, G. H. Jansen, S. Bulk, F. Baas and W. A. van Gool, *Neurology*, 2000, **55**, 514.
76. R. Siman, V. L. Roberts, E. McNeil, A. Dang, J. E. Bavaria, S. Ramchandren and M. McGarvey, *Brain Res.*, 2008, **1213**, 1.
77. R. Siman, T. K. McIntosh, K. M. Soltesz, Z. Chen, R. W. Neumar and V. L. Roberts, *Neurobiol Dis.*, 2004, **16**, 311.
78. R. Siman, C. Zhang, V. L. Roberts, A. Pitts-Kiefer and R. W. Neumar, *J. Cereb. Blood Flow Metab.*, 2005, **25**, 1433.
79. J. B. Redell, A. N. Moore, N. H. Ward, 3rd, G. W. Hergenroeder and P. K. Dash, *J. Neurotrauma*, 2010, **27**, 2147.
80. T. L. Barr, S. Alexander and Y. Conley, *Biol. Res. Nurs.*, 2011, **13**, 140.
81. Y. W. Zhou, Y. G. Zhang and W. N. Deng, *J. Forensic Med.*, 2002, **18**, 146.
82. D. B. Michael, D. M. Byers and L. N. Irwin, *J. Clin. Neurosci.*, 2005, **12**, 284.
83. J. Kukacka, D. Vajtr, D. Huska, R. Prusa, L. Houstava, F. Samal, V. Diopan, K. Kotaska and R. Kizek, *Neuro. Endocrinol. Lett.*, 2006, **27** (Suppl 2), 116.
84. T. Deierborg, L. Roybon, A. R. Inacio, J. Pesic and P. Brundin, *Neuroscience*, 2010, **171**, 1386.

CHAPTER 2

Magnetic Resonance Imaging Biomarkers of Mild Traumatic Brain Injury

ZHIFENG KOU, Ph.D.,*[a] RANDALL R. BENSON, M.D.,[b]
AND E. MARK HAACKE, Ph.D.[a]

[a] Departments of Biomedical Engineering and Radiology, Wayne
State University School of Medicine, Detroit, Michigan 49098 USA;
[b] Departments of Neurology and Radiology, Wayne State University
School of Medicine, Detroit, Michigan 48201, USA
*Email: zhifeng_kou@yahoo.com

2.1 Introduction

Traumatic brain injury (TBI) affects 1.7 million Americans each year,[1,2] the majority of which are mild TBI (mTBI).[3] Today, more people who might have had more severe TBI or died in motor vehicle accidents suffer milder head injury, thanks to the advancements in motor vehicle safety. mTBI affects over 1 million Americans annually. It is a major public healthcare burden, yet has been overlooked for decades.[1,2] Despite its name, the impact of mTBI on the patients and their family is not mild at all.[3] mTBI patients typically develop a variable constellation of physical, cognitive, and emotional symptoms, collectively known as post-concussive syndrome (PCS), that significantly impact their quality of life. The direct cost of mTBI in the U.S. is approximately $16.7 billion annually and this does not include the indirect costs to society and families.[3–5]

Up to 50% of patients with mTBI have persistent neurocognitive symptoms at three months and 5% to 15% at 1 year.[6,7] Meanwhile, among the 1.8 million

RSC Drug Discovery Series No. 24
Biomarkers for Traumatic Brain Injury
Edited by Svetlana A. Dambinova, Ronald L. Hayes and Kevin K. W. Wang
© Royal Society of Chemistry 2012
Published by the Royal Society of Chemistry, www.rsc.org

troops who have served in Iraq and Afghanistan, it is estimated that at least 20% of returning troops have suffered at least one, and some multiple, concussions. This means up to 360 000 veterans may have brain injuries after discharge.[8] However, most *symptomatic* mTBI patients have normal CT and conventional MRI scans,[9,10] even with the greater sensitivity of MRI to TBI.[11] In addition to neuroimaging, clinical indices of severity, such as the Glasgow Coma Scale (GCS) and duration of post-traumatic amnesia (PTA) are lacking in sensitivity in mTBI and are not helpful in predicting outcome.[12] In summary, using currently available clinical instruments, it is difficult to determine which mTBI patients will have prolonged or even permanent neurocognitive symptoms.

With greater appreciation of incidence and impact of "civilian," sports, and military mTBI, there is an increasing need to identify a minimally invasive biomarker (or biomarkers) of mTBI. A clinically useful biomarker or biomarkers should ideally be very sensitive to TBI (including mild cases) but should have reasonable specificity. That TBI pathology is not singular but involves multiple functional and structural changes affords greater opportunity for detection and characterization of injury with sensitive instruments. Similarly, biomarkers meeting these criteria should improve clinical trial efficiency since patients with similar types and grades of pathology could be grouped together in order to reduce the chances of a negative result due to prohibitive pathologic heterogeneity.

With both hardware and pulse sequence design advances, newer MRI methods have demonstrated the ability to detect and localize with high resolution *several* of the pathologic and pathophysiologic consequences of mTBI. These advanced MR technologies include susceptibility weighted imaging (SWI) for hemorrhage detection,[13] MR spectroscopy (MRS) for metabolite measurement,[14] diffusion weighted and diffusion tensor imaging (DWI/DTI) for edema quantification[15] and axonal injury detection,[16] perfusion weighted imaging (PWI) and arterial spin labeling (ASL) to measure blood flow to brain tissue,[17] and functional MRI (fMRI), which measures changes in blood oxygen level locally in response to neuronal activity.[18] Having a number of imaging biomarkers, all of which are obtained in a single scanning session (or multiple for longitudinal study) and are sensitive to different consequences of traumatic injury affords great advantages: 1) enhanced sensitivity; 2) ability to study inter-relationships among these biomarkers and between the biomarkers and clinical/neurocognitive deficits; 3) improved clinical management resulting from more precise characterization of injuries; and 4) enhanced power of clinical interventional studies (see above). Furthermore, the information comes "cheaply" since MRI is non-invasive and is widely available.

2.2   Search Criteria and Results

The purpose of this chapter is to review the current status and clinical impact of MR imaging findings in mTBI. BeLanger *et al.*[19] provided an informative review on this topic in 2007. Since then many original works have been published.

The following inclusion criteria were used to identify potential papers: a) must be cited in Medline database; b) must use MRI as a major component of the study; c) only clinical (human) studies of mTBI patients; d) published in the past five years (i.e. 2006 and later); and e) published in peer-reviewed English language journals. Readers interested in moderate to severe TBI imaging are encouraged to refer to another recent review paper, which covers SWI, DTI, and MRS.[19]

The authors conducted a Medline search at www.pubmed.com in September 2011 by using the search phrases "mild traumatic brain injury" and "magnetic resonance imaging." There were a total 356 publications, of which 46 were reviews and 310 were original studies. After excluding articles that did not fit our criteria, we found a total 62 articles; 11 reviews and 51 original studies. Of original studies, 32 used structural imaging, DTI, MRS, and perfusion imaging and 19 articles used fMRI, including both resting state fMRI and task-oriented fMRI. This review focuses on clinical MRI, DTI, MRS, and perfusion imaging. The following was categorized according to the technical approaches being used. The literature is listed in Table 2.1.

2.3 Clinical CT and MRI

Although CT is still the first step in the search for large hemorrhages that may require surgical intervention, it is insensitive to small hemorrhages, whether from early contusion or from diffuse axonal injury. Most patients with mTBI have normal CT and conventional MRI findings. Indeed, they usually do not get clinical MRI scans in the emergency department, which sees the majority of patients with mTBI. Any intracranial abnormality would classify the patients as complicated mTBI,[20] which could have long-term sequelae similar to moderate TBI. A clinical radiological sequence protocol typically includes T1-weighted imaging for structural damage and brain atrophy, T2-weighted FLAIR for edema and acute stage subarachnoid hemorrhage, and T2* gradient recalled echo (GRE) for hemorrhages.

It would be of clinical value to know: "How many patients with mTBI have positive MRI findings?" and "How these findings may account for their neurocognitive or functional symptoms?" Regarding the prevalence of positive MRI findings in mTBI, different groups reported different findings. After reviewing 182 adolescent patients, Fay et al.[21] reported 18% had positive MRI findings. Topal et al.[22] reported that of 40 patients with mTBI, 12.5% had positive MRI. In contrast, Lee and colleagues[23] reported 50% with positive CT and 75% with positive MRI in 36 mTBI patients with both loss of consciousness and post-traumatic amnesia. The discrepancy among the figures could be due to the different study populations and different patient inclusion criteria.

Regarding the relationship between MRI findings and patient outcome, Muller et al.[24] reported that GCS < 14, CT/MRI positive, and S100B > 0.14, together, predicted impaired cognitive performance both at baseline and after 6 months. Kurca et al.[25] reported that patients with mTBI with true traumatic MRI lesions are different neuropsychologically from those of mTBI with non-specific MR lesions, and neurocognitive and symptom signs have organic bases

Table 2.1 Summary of literature search result.

First author and year	Imaging approach	Number of patients	Inclusion criteria	Imaging stage	Analytical approach	Major findings
Clinical MRI						
Fay 2010[21]	Clinical MRI (T1, T2, and FLAIR)	182 children	LOC or GCS<15 or 2 + PCS symptoms	Acute	Clinical radiological diagnosis	18% with positive MRI finding (complicated mTBI); ratings of PCS were moderated by both cognitive ability and injury severity; children with lower cognitive ability with cmTBI were prone to PCS
Muller 2009[24]	Clinical MRI, serum S-100B, ApoE	59 mTBI			Clinical radiological diagnosis	GCS<14, CT/MRI+ and S100B>0.14 predicted impaired cognitive performance both at baseline and after 6 months; APoE allele predicted less recovery of cognitive function after mTBI
Lee 2008[23]	CT and clinical MRI	36 mTBI	Both LOC and PTA	Acute CT, subacute MRI (in 2 weeks)	Clinical radiological diagnosis	CT detected 50% with abnormality; MRI detected 75% with abnormalities. CT and MR finding did not account for cognitive impairment
Topel 2008[22]	CT and clinical MRI	40 mTBI	Normal CT, GCS 13–15	Acute (within 24 hours)	Clinical radiological diagnosis	12.5% patients had MR abnormality, 10% with hemorrhagic lesion on T2 GRE, and 12.5% with high intensity on FLAIR and DW sequences
Lewine 2007[101]	MEG, SPECT, and MRI	30 mTBI, retrospective review	GCS 13–15 and LOC <20 min	subacute (Weeks after TBI)	Clinical radiological diagnosis	MEG is more sensitive to brain injury and correlated with memory and attention problems
Kurca 2006[25]	MRI	30 mTBI	WHO definition: GCS 13–15, LOC, PTA or mental status change	Acute (≤4 days)	Clinical radiological diagnosis	mTBI patients with true traumatic MRI lesions are different neuropsychologically from mTBI with non-specific MR lesions; neurocognitive and symptom signs have real organic basis, which can be detected by MRI

Study	Technique	Sample	Criteria	Timing	Analysis	Findings
Inglese 2006[102]	MRI	38 mTBI		Acute and chronic	Clinical radiological diagnosis	Average number of Virchow Robin Space was significantly higher in mTBI patients than in controls. VRS was not associated with neurocognitive findings
Perfusion imaging						
Ge 2009[61]	ASL perfusion	21 mTBI	GCS 13–15, brief LOC, and PTA	Chronic	ROI on thalamus	Reduced CBF in both sides of thalamus, in correlation with speed of information processing, memory, verbal, and executive function
MR spectroscopy imaging						
Kirov 2007[85]	2D MRSI	20 mTBI		Chronic	Mixed model regression, LCModel	mTBI had more variations in NAA, Cre and Cho than controls
Cohen 2007[86]	MRS, MRI	$N = 20$ mTBI	GCS 13–15 and LOC	Acute and chronic	Single voxel MRS, T1 volumetry	Whole brain NAA decreased by 12%; mTBI patients had global and GM atrophy
Yeo 2011[103]	MRSI	$N = 30$ mTBI	ACRM	Subacute (within 21 days)	MRSI, LC model	Cre and Glx increase at WM and Glx decrease at GM at subacute stage; GM Glx normalized and WM Cre and Glx tend to normalize at 4 months after injury
Govind 2010[91]	3D MRSI	$N = 20$ mTBI, 9 moderate TBI	GCS 10–15	Subacute (mean 21 days)	3D MRSI	Widespread decrease of NAA and NAA/Cre, and increases of Cho and Cho/NAA in whole brain, and with the largest differences in WM; negative correlations between Cho/NAA with neurocognitive score in the frontal lobe; no significant correlations between any MRSI or neuropsychological measures and the GCS

Table 2.1 (*Continued*)

First author and year	Imaging approach	Number of patients	Inclusion criteria	Imaging stage	Analytical approach	Major findings
Diffusion imaging						
Grossman 2011[60]	Diffusion Kurtosis imaging (DKI)	22 mTBI	ACRM	Subacute to chronic	Diffusion Kurtosis analysis	Changed DTI and DKI measures in thalamus and IC, and CC and centrum semiovale in more chronic patients; cognitive impairment associated with mean kurtosis in thalamus and IC
MacDonald 2011[63]	DTI	$N = 63/21$, Service personnel	LRMC[119]	Subacute	ROI	Abnormalities in the middle cerebellar peduncles, cingulum bundles, and in the right orbitofrontal white matter
Cubon 2011[48]	DTI	10 college Sports concussion	No LOC, symptom based diagnosis	Chronic	TBSS	MD increase in left WM fibre tracts, IC, and thalamic acoustic radiations
Sponheim 2011[64]	EEG and DTI	9 blast-mTBI soldiers	ACRM definition	N/A	ROI of ICBM atlas	Diminished interhemispheric coordination of brain activity; EEG Phase synchrony associated with FA in frontal WM
Messe 2011[45]	DTI	23 mTBI	ACRM	Subacute	TBSS	MD increase associated with patients' outcome
Holli 2010[104,105]	Structural MRI, DTI	42 mTBI	WHO definition on mTBI[122]	Subacute, (3 weeks)	ROI Texture analysis of MRI, DTI ROI	Textural differences between left and right hemispheres; Texture parameters in mesencephalon and genu of cc correlate with memory; texture parameters in mesencephalon correlated with DTI FA and ADC
Mathew 2011[65]	fMRI, DTI	11 bTBI with major depression disorder	LOC or change of mental status during interview	Chronic	Face matching fMRI, DTI VBA	Hyperactivity in amygdala and other emotional processing structure; Low FA in SLF; depression symptoms correlated with FA in SLF

Study	Modality	Sample	ACRM	Time	Analysis	Findings
Mayer 2010[42]	DTI	22 mTBI		Subacute (12 days)	ICBM atlas ROI	Normal structural imaging, FA increase, reduced radial diffusivity in CC and left hemisphere tracts
Chu 2010[38]	DTI	10 adolescents	Normal CT, GCS 15, brief LOC	Acute (1–6 days post)	VBA	Increase FA, reduced ADC, and reduced radial diffusivity in WM regions and in left thalamus
Wu 2010[39]	DTI	12 adolescents	Brief LOC	Acute (1–6 days)	ROI	Lower ADC than controls; FA increase of left cingulum bundle correlated with delayed recall
Lipton 2009[46]	DTI	20 mTBI	LOC <20 min, PTA <24 hours, GCS ≤15, not neurologic deficit	2–24 days	VBA	FA reduced in WM of dorsolateral prefrontal cortex (DLPRC), correlated with worse executive function; MD increased in several clusters of WM; FA increase in the posterior limb of IC bilaterally
Lo 2009[49]	DTI	10 mTBI with persistent cognitive symptoms		Chronic	ROI	Reduced FA and increased ADC at the left side of CC genu. FA increased in IC bilaterally
Lipton 2008[82]	DTI	17 mTBI with persistent cognitive impairment	LOC <20 min PTA <24 hours, GCS ≤15, not neurologic deficit	Chronic	Whole brain histogram; VBA	Histogram of WM FA shift to lower end; Reduced FA and increased mean diffusivity in CC, subcortical WM, and IC bilaterally
Niogi 2008[62]	DTI	43 mTBI	LOC and PTA	Chronic	ROI	Reduced FA in left anterior corona radiata in correlation with attentional control, and reduced FA in uncinate fasciculus in correlation with memory performance
Wilde 2008[40]	DTI, SWI	10 adolescents	Normal CT, GCS = 15, brief LOC	Acute (1–6 days post)	ROI on CC	One patient with SWI bleed; increased FA and decreased ADC and MD in CC, in correlation with PCS score

Table 2.1 (*Continued*)

First author and year	Imaging approach	Number of patients	Inclusion criteria	Imaging stage	Analytical approach	Major findings
Niogi 2008[50]	DTI, T2* GRE	34 mTBI with persistent symptoms	GCS 13–15, LOC and PTA; 1 + symptom at time of imaging	Subacute to Chronic	ROI on major WM tracts	FA decreased in several WM regions: anterior corona radiata, uncinate fasciculus, cc genu, cingulum bundle; number of damaged WM regions correlated with reaction time.
Miles 2008[47]	DTI	17 mTBI		Subacute	ROI	Increased MD and reduced FA in mTBI patients; baseline MD tend to correlate with response speed.
Rutgers 2008[66]	DTI fibre tracking	21 mTBI	Clinical diagnosis	Chronic	VBA	Patients had multiple WM regions with reduced FA; 19.3% fiber bundles show discontinuity on fiber tracking
Bazarian 2007[41]	DTI	6 mTBI/6 orthopedic injury controls	ACRM	Acute (within 72 hours)	VBA and ROI	mTBI with lower mean trace in left anterior IC and higher FA in posterior IC. FA values correlated with 72-hour PCS score and neurobehavioral tests (visual motor speed and impulse control)
Inglese 2005[43]	DTI	N = 46 mTBI		20 acute and 26 chronic	Whole brain histogram, and ROI	Reduced FA in patients' CC, IC, and centrum semiovale. Increased MD in cc and IC. No group difference

Abbreviations: ACRM, American Congress of Rehabilitation Medicine; ADC, apparent diffusion coefficient; ASL, arterial spin labeling; CBF, cerebral blood flow; CC, corpus callosum; CT, computed tomography; DKI, diffusion kurtosis imaging; FA, fractional anisotropy; FLAIR, fluid attenuated inversion recovery; fMRI, functional MRI; GCS, Glasgow Coma Scale; GM, grey matter; IC, internal capsule; ICBM, International Consortium of Brain Mapping; LRMC, Landstuhl Regional Medical Centre; MD, mean diffusivity; MEG, magnetic encephalography; MRI, magnetic resonance imaging; PCS, post-concussive syndrome; PTA, post-traumatic amnesia; ROI, region of interest; SLF, superior longitudinal fasciculus; SPECT, single-photon emission computed tomography; TBSS, tract-based spatial statistics; THA, thalamus; VBA, voxel-based analysis; VRS, Virchow Robin Space; WHO, World Health Organization; WM, white matter.

that can be detected by MRI. Fay *et al.*[21] reported that ratings of PCS in adolescents were moderated by both cognitive ability and injury severity. Children with lower cognitive ability with complicated mTBI were prone to PCS. Comparatively, Lee *et al.*[23] reported that CT and MRI findings were not associated with a patient's neurocognitive impairment. Clearly, the seemingly conflicting findings on the predictive role of structural MRI need further investigation.

2.3.1 Susceptibility Weighted Imaging (SWI) of Hemorrhagic Lesions

In diagnostic radiology, intracranial hemorrhage has been sought as a biomarker of TBI. Some investigators have suggested that the presence of hemorrhage in DAI is predictive of poor outcome in moderate to severe TBI.[26] SWI was developed by Haacke *et al.*[27] as a high resolution venography method. It has been used to evaluate moderate to severe TBI patients since 2003 by Tong *et al.*,[13] who have shown that SWI is 3 to 6 times more sensitive than conventional T2* gradient echo imaging (GRE) for detecting suspected DAI lesions in children.[13,28] SWI has been shown to detect tiny hemorrhages that may be the only abnormal finding that can confirm the presence of brain injury, and change management of the patient. In addition, lesion number and volume identified by SWI are negatively associated with patient outcome[28] and neuropsychological functions[29] in patients with moderate to severe injury.

After brain injury, the hemorrhagic bleed may undergo a temporal transformation from oxyhemoglobin and then deoxyhemoglobin in the acute stage; intra- and then extracellular methemoglobin in the subacute stage; and finally, hemosiderin in the chronic stage.[30–32] SWI is sensitive to deoxyhemoglobin in the acute stage and extracellular methemoglobin in the subacute stage and hemosiderin in the chronic stage.[30] Therefore, positive results with SWI should provide a biomarker for hemorrhagic brain injury at any stage. However, very few data have been reported regarding the possible role of SWI in an improved detection of microhemorrhages and its predictive value of patient outcome in mTBI. Unlike DTI, which requires complex post-processing and comparison with proper controls, SWI hemorrhage is readily available for radiological reading immediately after MRI scans. This makes SWI more likely to have a direct impact on the radiological diagnosis. Furthermore, with quantitative susceptibility mapping, the hemorrhagic lesion load could be more acutely mapped through next generation SWI analysis and potentially improve the prediction of a patient's outcome. Figures 2.1 and 2.2 demonstrate two exemplar cases to show the SWI detection of brain hemorrhages in mTBI in the acute stage in comparison with the usual clinical T2* GRE or FLAIR.

In summary, there is still a need to evaluate a large number of patients with mTBI by using structural imaging, particularly SWI, to see the clinical value of intracranial hemorrhage detected by MRI. This should be done along with other clinical measures, e.g. GCS, as well as other biological markers, such as serum biomarkers.

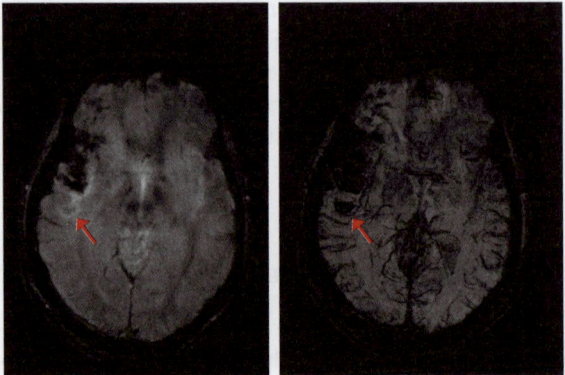

Figure 2.1 Comparison of T2* with SWI on the detection of hemorrhagic lesions. Image order: T2* GRE (left) and SWI (right). A 45-year-old man fell down a staircase and then visited ER with severe headache. Both T2* GRE and SWI detected hemorrhages on the right side of inferior temporal lobe. However, in a small area with mixed blood and edema signal, T2* GRE only detects edema (bright signal) and SWI detects hemorrhage (dark signal) (see arrows).

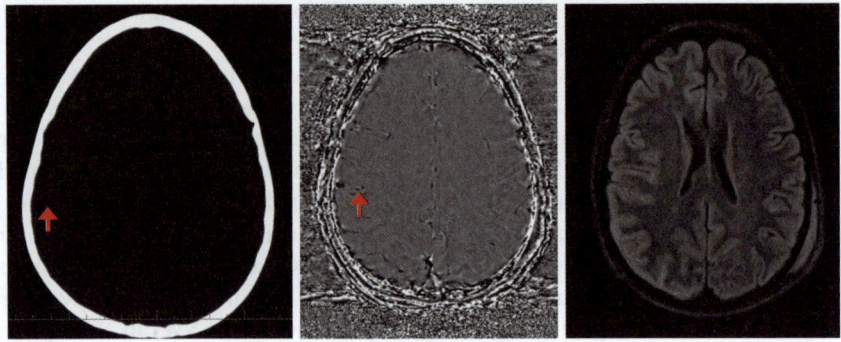

Figure 2.2 SWI detects subarachnoid hemorrhage (SAH) in an acute mTBI case. Image order from left to right: CT, SWI phase, and T2 FLAIR. A 21-year-old man fell from a 15-foot (4.5 m) high ladder. Both CT and SWI phase detected sub-arachnoid hemorrhage (see arrows) at the acute stage, while F2 FLAIR failed to detect.

2.4 Diffusion Tensor Imaging (DTI) of Axonal Injury

Diffusion imaging sequences are sensitive to traumatic axonal injury (TAI) secondary to stretch and shear forces. DTI measures the bulk motion of water molecular diffusion in biological tissues. It is most useful when tissues are anisotropic, i.e. when diffusion is not equivalent in all directions, such as in skeletal muscle or axons in white matter of the central nervous system. Histological correlates have validated DTI's sensitivity to brain injury for both focal[33] and diffuse axonal injury[34] models. The apparent diffusion coefficient (ADC) and fractional anisotropy (FA)[35,36] are two parameters derived from DTI that have been extensively studied in TBI. ADC is an estimate of the average *magnitude* of water movement in a voxel (regardless of direction), while FA is an index of the *directional non-uniformity*, or anisotropy, of water diffusion within a voxel.

FA has been used to detect alterations in directional diffusion resulting from tissue damage. FA in white matter is highest when fibers are long (relative to voxel dimension) and oriented uniformly (collinear) within a voxel and lowest when fibers are not collinear (e.g. "crossing fibers") or have been damaged. When axons are injured, as in acceleration/deceleration injuries (such as MVCs), diffusion anisotropy typically decreases. Loss of diffusion anisotropy is the result of a number of axonal changes after injury including: 1) increased permeability of the axonal membrane; 2) swelling of axons; 3) decreased diffusion in the axial (long axis) direction; 4) degeneration and loss of axons in the chronic stage. In general, any pathological alteration of white matter fibers will result in FA decrease, since one or more of these axonal changes occurs in disorders of white matter. Not surprisingly, most clinical studies of moderate and severe TBI have shown FA to be more sensitive than ADC to traumatic injury. On the other hand, ADC, FA, and directionally selective diffusivities (principal, intermediate, and minor components of diffusion) can help to better characterize brain injury pathologies. Trace and mean diffusivity are two other measurements similar to ADC and vary similarly. Changes in FA in association with ADC changes can differentiate the type of edema. For example, in the acute stage, decreased FA in association with increased ADC suggests vasogenic edema, while increased FA in association with decreased ADC suggests cytotoxic edema. Decreased FA in association with decreased ADC and decreased longitudinal (parallel to the long axis of the axons) water diffusivity suggests axonal transport failure as occurs in degenerative neurological diseases such as in ALS.

Regarding the location of brain lesions detected by DTI, Niogi *et al.*[37] summarized that the frontal association pathways, including anterior corona radiata, uncinate fasciculus, superior longitudinal fasciculus, and genu of corpus callosum (CC), were the mostly frequently injured WM structures in this cohort of mTBI patients. Single subject results were not reported but it can be assumed that significant inter-individual variability for location and extent of WM injury exists due to varying injury mechanisms (and forces) and biological (neural and non-neural) differences across patients.

2.4.1 Imaging at Different Pathological Stages

Interestingly, despite the higher incidence of milder TBI compared with more severe TBI in Western countries, there are many fewer mTBI imaging studies reported in the literature. One reason is the fact that recruitment of patients with mTBI is more difficult, since they are typically outpatients. Another reason is the common conception that mTBI is a transient problem from which virtually all who are afflicted will recover fully. Therefore, some question the clinical importance of studying mTBI. Certainly, insurance reimbursement for MRI scanning of mTBI is rare and even rarer in the acute setting, so that "adding on research sequences" is not possible. With these limitations in mind, a growing literature on DTI in TBI has begun to address the DTI findings at different stages.

2.4.1.1   Acute Stage

There are conflicting findings for FA and ADC in mTBI. Chu et al.,[38] Wu et al.,[39] and Wilde et al.[40] from Baylor College of Medicine scanned 10 to 12 adolescents with mTBI within 6 days of injury and reported *increased FA, reduced ADC* and *reduced radial (short axis) diffusivity* in WM regions and left thalamus. Similarly, Bazarian *et al.*[41] studied six mTBI patients within 72 hours and reported *increased FA* in the posterior CC and *reduced ADC* in the anterior limb of the internal capsule (IC). Similar to Chu *et al.*,[38] and Wilde *et al.*,[39] Mayer *et al.*[42] studied 22 patients with mTBI within 12 days of injury and reported *FA increase* and *reduced radial diffusivity* in the CC and left hemisphere tracts.

Inglese *et al.*,[43] in contradiction to Chu *et al.*,[38] Wilde *et al.*,[39,40] Bazarian *et al.*,[41] and Mayer *et al.*,[42] found *reduced FA* in the splenium of CC and posterior limb of IC in 20 patients with mTBI imaged up to 10 days after injury (mean = 4 days). Manually drawn regions of interest were used to assay the genu and splenium of the CC, the centrum semiovale, and the posterior limb IC bilaterally. In the same line as Inglese *et al.*, Arfanakis and colleagues[44] studied a handful of patients with mTBI at the acute stage and reported FA decrease in major WM tracts.

2.4.1.2   Subacute Stage

All studies reported *reduced FA* and/or *increased diffusivity*; i.e. ADC, trace or mean diffusivity (MD), at this stage. Messe *et al.*[45] studied 23 mTBI at the subacute stage and found significantly *increased MD* in mTBI patients with poor outcome.[45] The authors did not find significant changes of FA values. Lipton *et al.*[46] scanned 20 mTBI patients in the subacute stage and demonstrated *reduced FA* and *increased MD* in frontal subcortical WM. Miles *et al.*[47] studied 17 acute and subacute mTBI patients and found *reduced FA* and *increased MD*. All of these studies at the subacute stage reported a similar profile of DTI measures except for different locations of injury.

2.4.1.3   Chronic Stage

Three studies report either *reduced FA* or *increased diffusivity* or both. Cubon *et al.*[48] studied 10 collegiate athletes with concussion at the chronic stage and found MD increase in left WM tracts, internal capsule, and thalamic acoustic radiations. Lo *et al.* studied 10 mTBI patients and reported *reduced FA* and *increased ADC* in the left genu of the CC and *increased FA* in internal capsule bilaterally.[49] Niogi *et al.*[50] studied 43 patients with chronic mTBI and reported *reduced FA* in a large number of WM tracts.

All DTI studies of moderate to severe TBI patients[16,51–53] and subacute/ chronic mTBI patients[43,50,54–56] report FA *decreases*, which are correlated with clinical or neuropsychological measures. However, there are seemingly contradictory findings in mTBI in the acute stage (within 1 week after injury) in the literature: Inglese[43] and Arfanakis[44] both reported FA *decreases*, while Wilde[40] and Bazarian[41] reported FA *increases* and decreased radial diffusivity. It has been suggested that increased FA *acutely* may reflect cytotoxic edema,[40] which would shunt extracellular fluid into swollen cells. This could have the effect of reducing inter-axonal free water and therefore increasing anisotropy.

2.4.1.4   Longitudinal Studies

Only a few investigators have followed FA over time in the *same* patients. Sidaros *et al.*[57] studied 23 patients with severe TBI at 8 weeks and again at 12 months and found that partial recovery of initially depressed FA values in the internal capsule and centrum semiovale predicted a favorable outcome. Kumar *et al.*[58] studied 16 patients with moderate to severe TBI at 2 weeks or less, 6 months, and 2 years and found persistently reduced FA except in the genu of the corpus callosum, where there was partial normalization by 2 years. Recently, two studies by Mayer *et al.*[42] and Rutgers *et al.*[55] reported that FA may partially normalize, reflecting recovery. This evidence suggests that a systematic investigation of a large number of patients with mTBI at acute, subacute, and chronic stages is warranted to reveal the evolution of pathophysiology in mTBI.

2.4.2   Correlation Between DTI-Derived WM Injury Topography and Neuropsychological Deficits

Patients with mTBI often develop a constellation of physical, cognitive, and emotional symptoms that are collectively known as post-concussion syndrome (PCS). In terms of neurocognitive symptoms, there are four key domains implicated in chronic neuropsychological impairment after mTBI. These domains include: 1) higher-order attention, 2) executive function, 3) episodic memory, and 4) speed of information processing. To date, several studies have demonstrated typical mTBI cognitive profiles and association with DTI findings. Niogi *et al.* summarized the topographic and neurocognitive deficits.[37]

Damage to the frontal WM has been reported to be associated with impaired executive function. Lipton *et al.*[46] studied 20 acute to subacute patients and reported that reduced FA in WM of dorsolateral prefrontal cortex (DLPRC) is correlated with worse executive function. Frontal WM injury is also associated with attention deficit. Niogi *et al.*[50] reported that reduced FA in the left anterior corona radiata is correlated with attention control in chronic mTBI patients.

Injury at the temporal WM tracts or cingulum bundle may cause memory problems. Niogi *et al.* reported that reduced FA in uncinate fasciculus correlated with memory performance.[59] Wu *et al.*[39] reported that FA measure of left cingulum bundle correlated with delayed recall.

Injury of the callosal fibers has been reported to be associated with PCS scores. Wilde *et al.*[40] studied 10 adolescent patients with mTBI in the acute stage and reported that increased FA and decreased ADC and MD in corpus callosum is correlated with patient PCS score. Bazarian *et al.*[41] studied 6 patients with mTBI in the acute stage and reported a lower mean trace in the left anterior IC and a higher FA in the posterior CC. FA values correlated with patient's 72 hours PCS score and visual motor speed and impulse control.

The overall burden or extent of WM injury is associated with both speed of information processing and overall functional outcome. Niogi *et al.*[50] studied 34 subacute to chronic mTBI patients and reported that FA decreased in several WM regions, including anterior corona radiata, uncinate fasciculus, CC genu, and cingulum bundle. They demonstrated that the number of damaged WM regions is correlated with patient's reaction time. Miles *et al.*[47] studied 17 mTBI patients at the acute stage and followed them up to 6 months after injury. They reported that, at the acute stage, the increased mean diffusivity (MD) in centra semiovale, the genu and splenium of CC, and the posterior limb of IC tended to correlate with patient response speed at 6 months after injury. Regarding the overall outcome, Messe *et al.*[45] divided patients with mTBI into two outcome groups, poor outcome versus good outcome. Poor outcome patients showed significantly higher mean diffusivity (MD) values than both controls and good outcome patients in the corpus callosum, the right anterior thalamic radiations and the superior longitudinal fasciculus, the inferior longitudinal fasciculus and the fronto-occipital fasciculus bilaterally.

Interestingly, injury or reduced blood supply in the thalamus, which is the relay station of neuronal pathways, may cause a constellation of symptoms in speed of information processing, memory, verbal, and executive function. Grossman *et al.* studied 22 patients with subacute to chronic mTBI by using diffusion kurtosis imaging, which is a more advanced form of diffusion analysis, and demonstrated that overall cognitive impairment is associated with the diffusion measurement in the thalamus and internal capsule.[60] This work is along the same line of a perfusion study by the same group, which demonstrated that reduced blood flow in the thalamus correlated with patient's overall neurocognitive function.[61]

In summary, significant progress has been made by researchers in recent years regarding the prognostic value of DTI in the form of FA for neurocognitive outcome in patients with mTBI. There is still a need to evaluate the neural basis of a patient's recovery process, which requires longitudinal studies over a large number of patients.

2.4.3   Pros and Cons of Different Analytical Approaches

One challenge of DTI is post-processing and interpretation of the images. There are several approaches available, including whole brain histogram analysis, region of interest analysis, voxel-based analysis, and tract-based spatial statistics, to name a few. Each approach has pros and cons. There is no standardized procedure for DTI analysis; it is a moving target. The choice of analytical approach depends on the hypothesis and goal of the study. Table 2.2 lists the pros and cons of each approach.

A whole brain histogram analysis is used to segment out the whole brain WM and plot the FA frequency distribution.[16] Two studies have demonstrated

Table 2.2 Comparison of different DTI analytical approaches.

Approach	Pros	Cons
Whole brain histogram	Easy to implement	Not sensitive to subtle changes of mild brain injury; lack of regional injury information
Manual ROI	Easy to implement, detect regional injury	Inter-rater and intra-rater variability issue; time consuming
Automated or semi-auto-mated ROI	Objective, consistent, detect regional injury	Normalization may distort DTI indices; potential misregistration issue; small fraction of voxels may be averaged out
Voxel-based analysis	Data driven, automated, objective	Warping may cause DTI indices' change; misregistration cause artificial edging effect
Tract based spatial statistics	Data driven, automated, objective, least mis-registration issue.	Only look at the center line of WM tracts; may miss the target of lesions at other areas of WM

DTI, diffusion tensor imaging; ROI, region of interest; WM, white matter.

that moderate to severe TBI patients had their FA histograms shifted toward zero,[16,51] and the TBI group had larger variations than controls. In a mTBI study, Inglese *et al.*[43] did a whole brain histogram analysis by combining both grey and white matter, and did not find a group difference between mTBI patients and controls. Instead of combining both GM and WM, Lipton *et al.*[62] segmented out WM and demonstrated the mTBI patient group's FA histogram of WM only shifted to lower end. However, as their data demonstrated, it is not as sensitive as that of the voxel-based analysis. The advantage of the histogram analysis is that it is relatively easy to implement. The disadvantage is the lack of regional injury information and lack of sensitivity to subtle changes of the WM, particularly for mild cases. Unless the subject has widespread multifocal injury on WM tracts, a small fraction of WM injury may not demonstrate a shift in the histogram or change its kurtosis.

To test a hypothesis of injury at specific WM tract(s) in association with specific domain of neurocognitive or neural behavioral symptoms, researchers used region of interest (ROI) analysis by either manually drawing[41,43,44,47,49,50,59,63] or segmenting out the regions in a prior defined template in an automated or semi-automated way.[42,64] This could also provide needed lesion localization information, which should improve biomechanical models of TBI. Manually drawn ROI is relatively easy and convenient to do. One could use a color-encoded FA diffusion map as a guide to define the boundary of a target region. It works on the native space of the subject's brain without worrying about the distortion caused by the warping algorithm. However, it is time consuming and error prone. It requires the rater's prior knowledge of which specific tracts or locations to define the region. It also has a potential inter-rater and intra-rater variability problem.

An automated ROI analysis approach would need initially to transform the images into a standard reference space; for example, International Consortium of Brain Mapping (ICBM) atlas, by using SPM8, FSL, or other software, in order to derive statistical maps computed from comparing "cases" and non-TBI controls, and then further divide the WM into standardized regions that are already defined on the atlas. Then each regional FA, ADC, or Eigen values could be computed. In this category, several studies reported significant FA differences between mTBI patients and controls.[42,64] This method overcomes the subjective issue in manually drawn ROIs by using a computer algorithm to warp each subject to a standard space. The result is objective and consistent. However, the use of a computer algorithm for image warping and registration will introduce new problems. Due to individual variations, the warping of one image to a standard space could change the DTI indices' values from the native space. Furthermore, in a relatively large region like the splenium of corpus callosum, a small fraction of voxels with abnormal FA value could be averaged out by the surrounding voxels in the same region, rendering this method insensitive to small lesions.

Voxel-based analysis (VBA) overcomes the sensitivity issue of the ROI approach by comparing patient and control DTI images voxel by voxel to determine the abnormal voxels between groups (see Figure 2.3 for an example). This approach is fully automated, data-driven, and reproducible. Several studies reported different clusters of WM tracts with abnormal FA values by using this method.[38,41,46,62,65,66] However, it suffers from the same warping and misregistration problems as that of the automated ROI approach. Particularly at the edge of WM or GM/WM junction area, any misregistration will result in comparing WM of one group with GM of another group and result in abnormal values of DTI indices.

To overcome the misregistration problem, a relatively new approach, called tract-based spatial statistics (TBSS), has been introduced (see Figure 2.4 for an example).[67] First, it averages both patient and control brains to form a

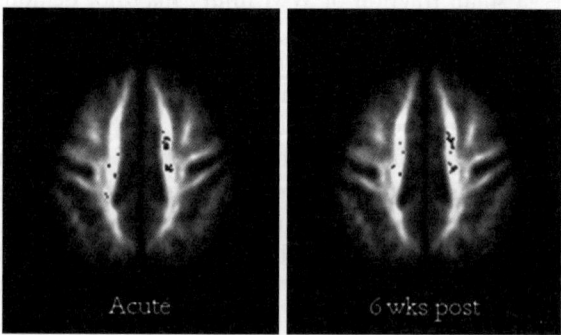

Figure 2.3   A mTBI patient scanned on the day of injury (5.5 hours after injury) and 6 weeks later. A 31-year-old male fell 10 feet (3.0 m) off a ladder, striking the back of his head with brief loss of consciousness and confusion. The patient developed persistent mild cognitive symptoms 6 weeks after injury. Note the same location of reduced FA in left corona radiata. Similar finding in splenium of CC (not shown). Global WM FA mean was within normal.

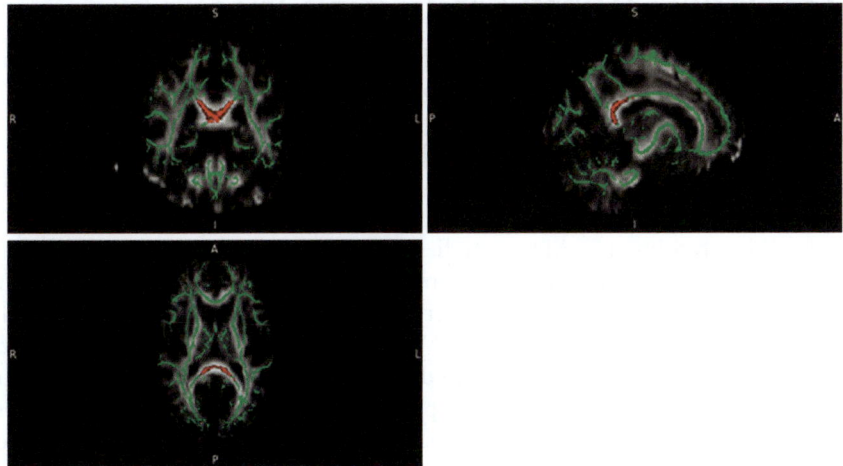

Figure 2.4 A demonstration of TBSS analysis to detect WM injury at the splenium of corpus callosum. A young man involved in a motor vehicle accident with negative structural MR images. TBSS analysis demonstrated significantly lower FA values at the splenium of corpus callosum, as demonstrated in the highlighted area. As shown on the image, it focuses on the skeleton (center lines) of WM.

template brain. Second, it defines a template skeleton of the WM tracts from the center lines of WM of the template brain. Third, all of the subjects' WM skeletons are nonlinearly registered onto the template skeleton. As such, the WM skeletons of all subjects will be aligned at the same space. Abnormal DTI indices could be defined along the skeletons by using a group comparison on a voxel-based analysis. Two groups reported the use of this method in the detection of mTBI[45,48] in association with patient outcome. This method is automated and reproducible and suitable for data-driven analysis. It has the least problem of misregistration. However, it only looks at the center lines of WM tracts. In thick WM bundles, like corpus callosum, it might miss lesions located off the center lines of WM.

2.4.4 DTI Caveat

In addition to the post-processing concerns mentioned above, another caveat of DTI is that it is experiment dependent. Different magnet field strength and image acquisition parameters, among others, will result in different DTI measures.[19] Furthermore, to date, most publications are performed with group comparisons between patients and controls instead of an individual-based analysis. However, radiological diagnosis is case-based. A radiologist or a clinician needs to make a decision for each individual subject instead of group comparison. The low sensitivity of DTI to individual case hinders its further implementation as a routine clinical sequence. A relatively standardized protocol is needed to test a large number of controls to determine the normal variations before deciding which specific case has brain injury.

2.5   *In Vivo* ^{1}H MR Spectroscopic Imaging

Cellular changes to neurons and glia following TBI are complex and dynamic. Proton (^{1}H) MRSI has the advantage of measuring brain metabolism *in vivo*, and is able to detect various biochemical processes of brain injury such as the loss (or dysfunction) of neuronal cells. TBI is known to induce changes in *N*-acetylaspartate (NAA) (a neuronal marker), choline (Cho) (a marker for membrane disruption, synthesis, or repair), and lactate (Lac) (a measure of anaerobic glycolysis).[68] Many studies of metabolic changes after TBI have proven MRS to be a sensitive tool to predict neurologic,[69–74] cognitive,[75–79] and functional outcome.[80] Significant decreases of NAA and increases of Cho have been observed in "normal appearing" WM or GM following moderate to severe TBI.[74–76,81–83] Elevated Cho may be detected in WM as a breakdown product after shearing of myelin[84] and the presence of lactate in the brain suggests hypoxic/ischemic injury. Researchers from Loma Linda University Medical Center found that increased frontal white matter (FWM) Cho/creatine and decreased FWM NAA/Cho at acute stage are useful for predicting functional outcome in patients with moderate to severe TBI. In addition, they also found that elevated Glx and Cho in the acute stage are sensitive indicators of injury and predictors of poor outcome in *normal appearing brain* of severe TBI patients.[71] They also found that the decrease of NAA at subacute stage is more predictive of patient outcome.

In mTBI studies, there are relatively less data reported in comparison with moderate to severe TBI studies. For a chronic mTBI study, Kirov *et al.*[85] scanned 20 mTBI patients and 17 age- and gender-matched controls and reported that the mTBI group had more variations in NAA, Cre, and Cho than the control group. Cohen *et al.*[86] investigated 20 patients with mTBI from 9 days to years after injury and found that whole brain NAA decreased by 12% along with significant global and grey matter atrophy, in comparison with 19 controls. This further demonstrates NAA as a biomarker of neuronal injury or cellular loss. Regarding the investigations of mTBI in the subacute or semi-acute stages, a few studies reported an NAA/Cr decrease,[86–88] which is suggestive of neuronal loss.

A recent study on 10 patients with mTBI in the semi-acute stage showed creatine increases in white matter and Glx (glutamate and glutamine) decreases in grey matter. The creatine (Cre) increases in white matter were correlated with patient's executive function.[89] The same research group also studied a relatively larger group of patients with longitudinal follow-up. In their study of 30 mTBI patients at the first visit within 3 weeks after injury, they reported Cre and Glx increases in WM and Glx decreases in GM.[90] Among the patients, 17 returned for follow-up at the chronic stage; their GM Glx demonstrated normalization and their WM Cre and Glx demonstrated a trend of normalization. Of particular note is that, at the first visit, the mTBI group did not have any difference from the controls in neuropsychological assessment except for more somatic, cognitive, and emotional symptoms.[90] This demonstrates that MRS may be sensitive to subtle changes of the brain, which may not yet be manifesting significant neural cognitive deficits.

As with other MRI techniques, MR spectroscopy has also undergone tremendous advancements in the past two decades, evolving from single voxel MR spectroscopy to 2D spectroscopy imaging and 3D whole brain spectroscopy imaging. Single voxel spectroscopy (SVS) allows acquisition of a single spectrum from one volume element (voxel) typically 8 cm^3 or more, whereas 2D or 3D magnetic resonance spectroscopic imaging (2D-MRSI/3D-MRSI), also called chemical shift imaging (CSI), allows for the simultaneous acquisition of multiple spectra from smaller adjacent voxels through multiple sections of the brain. MRSI has an inherent advantage over SVS because it is better able to evaluate regional distributions of neurochemical alterations. Instead of hypothesizing a certain region or certain slice of the brain to have abnormal metabolite signals; 3D MRSI allows investigators to search the whole brain to identify any area that might have abnormal metabolites. This could be useful in mTBI due to the fact that choosing an arbitrary region to study would be like searching for a needle in a haystack. Govind *et al.*[91] used a 3D-MRSI sequence to scan 20 mild and 9 moderate TBI patients and found a widespread decrease of NAA and NAA/Cre, and increases of Cho and Cho/NAA, within all lobes of the TBI subject group, and with the largest differences seen in WM. Examination of the association between all of the metabolite measures and the neuropsychological test scores found the strongest negative correlations to occur in the frontal lobe and for Cho/NAA.

One caveat of the previous MRS work on TBI is that Cre has been used as an internal reference with the assumption that Cre levels will remain constant after injury.[9] However, this might not be true in an injured brain, which could have Cre disturbance after TBI. As demonstrated by Gasparovic *et al.*,[89,90] the Cre increase in mTBI patients at the semi-acute stage might give a false impression of decreased NAA/Cre level. It is therefore important to quantitate metabolite concentrations carefully, particularly after mTBI, to determine possible small changes in all metabolites. Comparing metabolite ratios may mask important metabolite changes.

2.6 Perfusion Imaging

After the primary brain damage occurs, even more devastating secondary brain injuries can occur.[92] This usually includes, but is not limited to, ischemic and hypoxic damage, cerebral swelling, rise of intracranial pressure, hydrocephalus, and infection[93] in moderate to severe TBI patients at the acute stage. The injury to the brain can result in disturbance of cerebral blood flow (CBF), which correlates with poor brain tissue oxygenation,[94] and unfavorable neurological outcome,[95] which is implicated in rendering the brain vulnerable to secondary damage.[95] It has been recognized that the failure of brain perfusion is the most common type of hypoxic-ischemic brain damage, but is difficult to identify even in histological studies.[93] Cerebral ischemia (CBF ≤ 18 mL/100 g/min) is present in approximately 90% of patients who die with head injury.[92,96,97] Furthermore, after TBI, the uncoupling between CBF and brain metabolic demand occurs, whereby local cerebral glucose demand increases and local CBF is disturbed due to damage to the vascular muscle tone and arterioles.[98] The early hyper-glycolysis is thought to be due to the cellular efforts to reestablish ionic gradients.[95] Different aspects of ischemia and/or edema after injury can be quantified using perfusion imaging, including perfusion weighted imaging (PWI) and arterial spin labeling (ASL), and DWI/DTI. Kim *et al.*[17] used ASL to measure the resting state CBF in 28 moderate to severe TBI at chronic stage and found a) a global decrease of CBF in TBI patients than that in controls, and b) prominent regional hypoperfusion in the posterior cingulate cortices, the thalami, and multiple locations in the frontal cortices.

However, in the population of mTBI, how cerebral perfusion is being affected in patients and their association with patients' neurocognitive status is still less understood. Given the milder severity in mTBI patients, perfusion imaging at the acute stage could be meaningful both for clinical decision-making and patients' outcome prediction. Ge *et al.*[61] also employed ASL to study 21 mTBI patients at the chronic stage and demonstrated reduced CBF in both sides of the thalamus, which is correlated with the patients' speed of information processing, memory, verbal, and executive function.

In summary, our knowledge is very limited in the understanding of cerebral blood flow and its impact to mTBI patients. Perfusion imaging of mTBI, particularly at the acute stage, might shed new light in our understanding of this disease.

2.7   Delivering Improved Care to Patients with mTBI

Despite the fact that the emergency department (ED) sees the majority of patients with mTBI,[99] most stay only a few hours and then are discharged home. After that, most patients fail to be followed up. The acute stage, within 24 hours after injury, is the critical time point for imaging to deliver a real impact to medicine.[100] Either improved detection or outcome prediction will greatly help emergency physicians develop a better referral pattern or management plan for the patient or notify the patient's family members to arrange necessary resources in advance. However, most of the patients with mTBI at the acute stage do not get an MRI scan due to the high cost for medical insurance and the lack of accessibility to an MRI magnet. To date, very few studies are designed to target this critical time point. There is an urgent need for a comprehensive use of advanced MRI techniques to eval-uate the patients from the acute setting (within 24 hours) to the chronic stage after injury to identify the association between injury pathology and out-come. In the future, with the availability of MR magnets in the ED and reduced costs for an MRI scan, routinely scanning mTBI patients at the acute stage will be possible. However, to determine who will get an MRI scan in ED will be another issue, due to cost. In conjunction with other biochemical markers, a policy on clinical decision making would be expected to define a subgroup of patients for MRI scan.

2.8 Conclusions

MRI has demonstrated superior capabilities over CT in the detection of subtle changes of the brain after mTBI. As an advanced MRI technique, DTI can detect white matter abnormalities that are unseen in structural MRI and further map it with a patient's specific domain of neuropsychological symptoms. Other advanced MRI techniques also demonstrate the potential to detect other types of brain pathologies, including SWI for microhemorrhage detection, MRS for measuring abnormal cerebral metabolites, and perfusion imaging for detecting cerebral blood flow disturbances. In the future, these advanced MRI techniques should be used in a comprehensive way in a large cohort of patients to provide a panorama view of brain pathologies. The MRI investigations at the acute stage, within 24 hours after injury, will most likely impact emergency medicine, which is at the forefront of mTBI care. Furthermore, conjunctional use of imaging with other biological markers, such as serum biomarkers, will likely help screen patients who really need MRI in the acute setting.

References

1. T. Kay, *J. Head Trauma Rehabil.*, 1993, **8**, 74.
2. X. L. Faul, M. M. Wald and V. Coronado, *Traumatic Brain Injury in the United States: Emergency Department Visits, Hospitalizations, and Deaths, 2002–2006*, CDC, National Center for Injury Prevention and Control, Atlanta, GA, 2010.
3. CDC, *Report to Congress on Mild Traumatic Brain Injury in the United States: Steps to Prevent a Serious Public Health Problem*, Centers for Disease Control and Prevention, National Center for Injury Prevention and Control, Atlanta (GA), 2003.
4. J. J. Bazarian, J. McClung, M. N. Shah, Y. T. Cheng, W. Flesher and J. Kraus, *Brain Injury*, 2005, **19**, 85–91.
5. R. Ruff, *J. Head Trauma Rehabil.*, 2005, **20**, 5.
6. W. Alves, S. N. Macciocchi and J. T. Barth, *J. Head Trauma Rehabil.*, 1993, **8**, 48.
7. J. J. Bazarian, T. Wong, M. Harris, N. Leahey, S. Mookerjee and M. Dombovy, *Brain Inj.*, 1999, **13**, 173.
8. D. Warden, *J. Head Trauma Rehabil.*, 2006, **21**, 398.
9. H. G. Belanger, R. D. Vanderploeg, G. Curtiss and D. L. Warden, *J. Neuropsychiatry Clin. Neurosci.*, 2007, **19**, 5.
10. National Academy of Neuropsychology, *Mild Traumatic Brain Injury – An Online Course*, National Academy of Neuropsychology, Denver (CO), 2002.
11. E. Teasdale and D. M. Hadley, in *Head Injury*, ed. P. Reilly and R. Bullock, Chapman & Hall, London, 1997, p. 167.
12. A. Tellier, L. C. Della Malva, A. Cwinn, S. Grahovac, W. Morrish and M. Brennan-Barnes, *Brain Inj.*, 1999, **13**, 463.

13. K. A. Tong, S. Ashwal, B. A. Holshouser, L. A. Shutter, G. Herigault, E. M. Haacke and D. Kido, *Radiology*, 2003, **27**, 332.
14. B. A. Holshouser, K. A. Tong, S. Ashwal, U. Oyoyo, M. Ghamsary, D. Saunders and L. Shutter, *J. Magn. Reson. Imaging*, 2006, **24**, 33.
15. A. Marmarou, S. Signoretti, P. P. Fatouros, G. Portella, G. A. Aygok and M. R. Bullock, *J. Neurosurg.*, 2006, **104**, 720.
16. R. R. Benson, S. A. Meda, S. Vasudevan, Z. Kou, K. A. Govindarajan, R. A. Hanks, S. R. Millis, M. Makki, Z. Latif, W. Coplin, J. Meythaler and E. M. Haacke, *J. Neurotrauma*, 2007, **24**, 446.
17. J. Kim, J. Whyte, S. Patel, B. Avants, E. Europa, J. Wang, J. Slattery, J. C. Gee, H. B. Coslett and J. A. Detre, *J. Neurotrauma*, 2010, **27**, 1399.
18. T. W. McAllister, A. J. Saykin, L. A. Flashman, M. B. Sparling, S. C. Johnson, S. J. Guerin, A. C. Mamourian, J. B. Weaver and N. Yanofsky, *Neurology*, 1999, **53**, 1300.
19. Z. Kou, Z. Wu, K. A. Tong, B. A. Holshouser, R. R. Benson, J. Hu and E. M. Haacke, *J. Head Trauma Rehab.*, 2010, **25**, 267.
20. S. Margulies and R. Hicks, *J. Neurotrauma*, 2009, **26**, 925.
21. T. B. Fay, K. O. Yeates, H. G. Taylor, B. Bangert, A. Dietrich, K. E. Nuss, J. Rusin and M. Wright, *J. Int. Neuropsychol. Soc.*, 2010, **16**, 94.
22. N. B. Topal, B. Hakyemez, C. Erdogan, M. Bulut, O. Koksal, S. Akkose, S. Dogan, M. Parlak, H. Ozguc and E. Korfali, *Neurol. Res.*, 2008, **30**, 974.
23. H. Lee, M. Wintermark, A. D. Gean, J. Ghajar, G. T. Manley and P. Mukherjee, *J. Neurotrauma*, 2008, **25**, 1049.
24. K. Müller, T. Ingebrigtsen, T. Wilsgaard, G. Wikran, T. Fagerheim, B. Romner and K. Waterloo, *Neurosurgery*, 2009, **64**, 698.
25. E. Kurca, S. Sivak and P. Kucera, *Neuroradiology*, 2006, **48**, 661.
26. K. Paterakis, A. H. Karantanas, A. Komnos and Z Volikas, *J. Trauma*, 2000, **49**, 1071.
27. J. R. Reichenbach, R. Venkatesan, D. J. Schillinger, D. K. Kido and E. M. Haacke, *Radiology*, 1997, **204**, 272.
28. K. A. Tong, S. Ashwal, B. A. Holshouser, J. P. Nickerson, C. J. Wall, L. A. Shutter, R. J. Osterdock, E. M. Haacke and D. Kido, *Ann. Neurol.*, 2004, **56**, 36.
29. T. Babikian, M. C. Freier, K. A. Tong, J. P. Nickerson, C. J. Wall, B. A. Holshouser, T. Burley, M. L. Riggs and S. Ashwal, *Pediatr. Neurol.*, 2005, **33**, 184.
30. Z. Kou, R. R. Benson and E. M. Haacke, in *Clinical MR Neuroimaging*, 2nd edn, ed. J. Gillard, A. Waldman and P. Barker, Cambridge University Press, Cambridge, 2010.
31. K. R. Thulborn, A. G. Sorensen, N. W. Kowall, A. McKee, A. Lai, R. C. McKinstry, J. Moore, B. R. Rosen and T. J. Brady, *AJNR Am. J. Neuroradiol.*, 1990, **11**, 291.
32. W. G. Bradley, *Radiology*, 1993, **189**, 15.
33. C. L. Mac Donald, K. Dikranian, S. K. Song, P. V. Bayly, D. M. Holtzman and D. L. Brody, *Exp. Neurol.*, 2007, **205**, 116.

34. Z. Kou, Y. Shen, N. Zakaria, S. Kallakuri, J. M. Cavanaugh, Y. Yu, J. Hu and E. M. Haacke, in *Joint Annual Meeting ISMRM-ESMRMB*, Berlin, Germany, 2007, Abstract 824, p. 186.
35. J. S. Shimony, R. C. McKinstry, E. Akbudak, J. A. Aronovitz, A. Z. Snyder, N. F. Lori, T. S. Cull and T. E. Conturo, *Radiology*, 1999, **212**, 770.
36. T. E. Conturo, R. C. McKinstry, E. Akbudak and B. H. Robinson, *Mag. Reson. Med.*, 1996, **35**, 399.
37. S. N. Niogi and P. Mukherjee, *J. Head Trauma Rehabil.*, 2010, **25**, 241.
38. Z. Chu, E. A. Wilde, J. V. Hunter, S. R. McCauley, E. D. Bigler, M. Troyanskaya, R. Yallampalli, J. M. Chia and H. S. Levin, *AJNR Am. J. Neuroradiol.*, 2010, **31**, 340.
39. T. C. Wu, E. A. Wilde, E. D. Bigler, R. Yallampalli, S. R. McCauley, M. Troyanskaya, Z. Chu, X. Li, G. Hanten, J. V. Hunter and H. S. Levin, *J. Neurotrauma*, 2010, **27**, 361.
40. E. A. Wilde, S. R. McCauley, J. V. Hunter, E. D. Bigler, Z. Chu, Z. J. Wang, G. R. Hanten, M. Troyanskaya, R. Yallampalli, X Li, J. Chia and H. S. Levin, *Neurology*, 2008, **70**, 948.
41. J. J. Bazarian, J. Zhong, B. Blyth, T. Zhu, V. Kavcic and D. Peterson, *J. Neurotrauma*, 2007, **24**, 1447.
42. A. R. Mayer, J. Ling, M. V. Mannell, C. Gasparovic, J. P. Phillips, D. Doezema, R. Reichard and R. A. Yeo, *Neurology*, 2010, **74**, 643.
43. M. Inglese, S. Makani, G. Johnson, B. A. Cohen, J. A. Silver, O. Gonen and R. I. Grossman, *J. Neurosurg.*, 2005, **103**, 298.
44. K. Arfanakis, V. M. Haughton, J. D. Carew, B. P. Rogers, R. J. Dempsey and M. E. Meyerand, *Am. J. Neuroradiol.*, 2002, **23**, 794.
45. A. Messé, S. Caplain, G. Paradot, D. Garrigue, J. F. Mineo, G. Soto Ares, D. Ducreux, F. Vignaud, G. Rozec, H. Desal, M. Pélégrini-Issac, M. Montreuil, H. Benali and S. Lehéricy, *Human Brain Mapping*, 2011, **32**, 999.
46. M. L. Lipton, E. Gulko, M. E. Zimmerman, B. W. Friedman, M. Kim, E. Gellella, T. Gold, K. Shifteh, B. A. Ardekani and C. A. Branch, *Radiology*, 2009, **252**, 816.
47. L. Miles, R. I. Grossman, G. Johnson, J. S. Babb, L. Diller and M. Inglese, *Brain Inj.*, 2008, **22**, 115.
48. V. A. Cubon, M. Putukian, C. Boyer and A. Dettwiler, *J. Neurotrauma*, 2011, **28**, 189.
49. C. Lo, K. Shifteh, T. Gold, J. A. Bello and M. L. Lipton, *J. Comput. Assist. Tomogr.*, 2009, **33**, 293.
50. S. N. Niogi, P. Mukherjee, J. Ghajar, C. Johnson, R. A. Kolster, R. Sarkar, H. Lee, M. Meeker, R. D. Zimmerman, G. T. Manley and B. D. McCandliss, *AJNR Am. J. Neuroradiol.*, 2008, **29**, 967.
51. V. F. Newcombe, G. B. Williams, J. Nortje, P. G. Bradley, S. G. Harding, P. Smielewski, J. P. Coles, B. Maiya, J. H. Gillard, P. J. Hutchinson, J. D. Pickard, T. A. Carpenter and D. K. Menon, *Br. J. Neurosurg.*, 2007, **21**, 340.

52. H. S. Levin, E. A. Wilde, Z. Chu, R. Yallampalli, G. R. Hanten, X. Li, J. Chia, A. C. Vasquez and J. V. Hunter, *J. Head Trauma Rehabil.*, 2008, **23**, 197.

53. Z. Kou, R. Gattu, R. R. Benson, N. Raz and E. M. Haacke, *Proc. Intl. Soc. Mag. Reson. Med.*, 2008, **16**, 2272.

54. J. R. Wozniak, L. Krach, E. Ward, B. A. Mueller, R. Muetzel, S. Schnoebelen, A. Kiragu and K. O. Lim, *Arch. Clin. Neuropsychol.*, 2007, **22**, 555.

55. D. R. Rutgers, P. Fillard, G. Paradot, M. Tadié, P. Lasjaunias and D. Ducreux, *AJNR Am. J. Neuroradiol.*, 2008, **29**, 1730.

56. M. F. Kraus, T. Susmaras, B. P. Caughlin, C. J. Walker, J. A. Sweeney and D. M. Little, *Brain*, 2007, **130**, 2508.

57. A. Sidaros, A. W. Engberg, K. Sidaros, M. G. Liptrot, M. Herning, P. Petersen, O. B. Paulson, T. L. Jernigan and E. Rostrup, *Brain*, 2008, **131**, 559.

58. R. Kumar, S. Saksena, M. Hussain, A. Srivastava, R. K. Rathore, S. Agarwal and R. K. Gupta, *J. Head Trauma Rehabil.*, 2010, **25**, 31.

59. S. N. Niogi, P. Mukherjee, C. E. Ghajar, C. E. Johnson, R. Kolster, H. Lee, M. Suh, R. D. Zimmerman, G. T. Manley and B. D McCandliss, *Brain*, 2008, **131**, 3209.

60. E. J. Grossman, Y. Ge, J. H. Jensen, J. S. Babb, L. Miles, J. Reaume, J. M. Silver, R. I. Grossman and M Inglese, *J. Neurotrauma*, 2011, [Epub ahead of print].

61. Y. Ge, M. B. Patel, Q. Chen, E. J. Grossman, K. Zhang, L. Miles, J. S. Babb, J. Reaume and R. I. Grossman, *Brain Inj.*, 2009, **23**, 666.

62. M. L. Lipton, E. Gellella, C. Lo, T. Gold, B. A. Ardekani, K. Shifteh, J. A. Bello and C. A. Branch, *J. Neurotrauma*, 2008, **25**, 1335.

63. C. L. MacDonald, A. M. Johnson, D. Cooper, E. C. Nelson, N. J. Werner, J. S. Shimony, A. Z. Snyder, M. E. Raichle, J. R. Witherow, R. Fang, S. F. Flaherty and D. L. Brody, *N. Engl. J. Med.*, 2011, **364**, 2091.

64. S. R. Sponheim, K. A. McGuire, S. S. Kang, N. D. Davenport, S. Aviyente, E. M. Bernat and K. O. Lim, *Neuroimage*, 2011, **54**, S21.

65. S. C. Matthews, I. A. Strigo, A. N. Simmons, R. M. O'Connell, L. E. Reinhardt and S. A. Moseley, *Neuroimage*, 2011, **54**, S69.

66. D. R. Rutgers, F. Toulgoat, J. Cazejust, P. Fillard, P. Lasjaunias and D. Ducreux, *AJNR Am. J. Neuroradiol.*, 2008, **29**, 514.

67. S. M. Smith, M. Jenkinson, H. Johansen-Berg, D. Rueckert, T. E. Nichols, C. E. Mackay, K. E. Watkins, O. Ciccarelli, M. Z. Cader, P. M. Matthews and T. E. Behrens, *Neuroimage*, 2006, **31**, 1487.

68. B. Alessandri, E. Doppenberg, A. Zauner, J. Woodward, S. Choi and R. Bullock, *Acta Neurochir. Suppl. (Wien)*, 1999, **75**, 25.

69. B. A. Holshouser, S. Ashwal, G. Y. Luh, S. Shu, S. Kahlon, K. L. Auld, L. G. Tomasi, R. M. Perkin and D. B. Hinshaw, Jr., *Radiology*, 1997, **202**, 487.

70. B. A. Holshouser, S. Ashwal, S. Shu and D. B. Hinshaw, Jr., *J. Magn. Reson. Imaging*, 2000, **11**, 9.

71. L. Shutter, K. A. Tong and B. A. Holshouser, *J. Neurotrauma*, 2004, **21**, 1693.
72. B. A. Holshouser, K. A. Tong, S. Ashwal, U. Oyoyo, M. Ghamsary, D. Saunders and L. Shutter, *J. Magn. Reson. Imaging*, 2006, **24**, 33.
73. M. R. Garnett, A. M. Blamire, B. Rajagopalan, P. Styles and T. A. D. Cadoux-Hudson, *Brain*, 2000, **123**, 1403.
74. M. R. Garnett, R. G. Corkill, A. M. Blamire, B. Rajagopalan, D. N. Manners, J. D. Young, P. Styles and T. A. D. Cadoux-Hudson, *J. Neurotrauma*, 2001, **18**, 231.
75. W. M. Brooks, C. A. Stidley, H. Petropoulos, R. E. Jung, D. C. Weers, S. D. Friedman, M. A. Barlow, W. L. Sibbett and R. A. Yeo, *J. Neurotrauma*, 2000, **17**, 629.
76. S. D. Friedman, W. M. Brooks, R. E. Jung, B. L. Hart and R. A. Yeo, *AJNR Am J Neuroradiol.*, 1998, **19**, 1879.
77. T. Brenner, M. C. Freier, B. A. Holshouser, T. Burley and S. Ashwal, *Pediatr. Neurol.*, 2003, **28**, 104.
78. T. Babikian, M. C. Freier, S. Ashwal, M. L. Riggs, T. Burley and B. A. Holshouser, *J. Magn. Reson. Imag.*, 2006, **24**, 801.
79. R. A. Yeo, J. P. Phillips, R. E. Jung, A. J. Brown, R. C. Campbell and W. M. Brooks, *J. Neurotrauma*, 2006, **23**, 1427.
80. S. J. Yoon, J. H. Lee, S. T. Kim and M. H. Chun, *Clin. Rehabil.*, 2005, **19**, 209.
81. S. Signoretti, A. Marmarou, P. Fatouros, R. Hoyle, A. Beaumont, S. Sawauchi, R. Bullock and H. Young, *Acta Neuochir. Suppl.*, 2002, **81**, 373.
82. M. R. Garnett, A. M. Blamire, R. G. Corkill, T. A. D. Cadoux-Hudson, B. Rajagopalan and P. Styles, *Brain*, 2000, **123**, 2046.
83. S. D. Friedman, W. M. Brooks, R. E. Jung, S. J. Chiulli, J. H. Sloan, B. T. Montoya, B. L. Hart and R. A. Yeo, *Neurology*, 1999, **52**, 1384.
84. B. D. Ross, T. Ernst, R. Kreis, L. J. Haseler, S. Bayer, E. Danielsen, S. Bluml, T. Shonk, J. C. Mandigo, W. Caton, C. Clark, S. W. Jensen, N. L. Lehman, E. Arcinue, R. Pudenz and C. H. Shelden, *J. Magn. Reson. Imag.*, 1998, **8**, 829.
85. I. Kirov, L. Fleysher, J. S. Babb, J. M. Silver, R. I. Grossman and O. Gonen, *Brain Inj.*, 2007, **21**, 1147.
86. B. A. Cohen, M. Inglese, H. Rusinek, J. S. Babb, R. I. Grossman and O. Gonen, *AJNR Am. J. Neuroradiol.*, 2007, **28**, 907.
87. V. Govindaraju, G. E. Gauger, G. T. Manley, A. Ebel, M. Meeker and A. A. Maudsley, *AJNR Am. J. Neuroradiol.*, 2004, **25**, 730.
88. M. R. Garnett, A. M. Blamire, B. Rajagopalan, P. Styles and T. A. D. Cadoux-Hudson, *Brain*, 2000, **123**, 1403.
89. C. Gasparovic, R. Yeo, M. Mannell, J. Ling, R. Elgie, J. Phillips, D. Doezema and A. R. Mayer, *J. Neurotrauma*, 2009, **26**, 1635.
90. R. A. Yeo, C. Gasparovic, F. Merideth, D. Ruhl, D. Doezema and A. R. Mayer, *J. Neurotrauma*, 2011, **28**, 1.
91. V. Govind, S. Gold, K. Kaliannan, G. Saigal, S. Falcone, K. L. Arheart, L. Harris, J. Jagid and A. A. Maudsley, *J. Neurotrauma*, 2010, **27**, 483.

92. A. D. Mendelow and P. J. Crawford, in *Head Injury*, ed. P. Riley and R. Bullock, Chapman & Hall, London, 1997, p. 71.
93. P. C. Blumbergs, in *Head Injury*, ed. P. Reilly and R. Bullock, Chapman & Hill, London, 1997, p. 39.
94. A. Zauner, W. P. Daugherty, M. R. Bullock and D. S. Warner, *Neurosurgery*, 2002, **51**, 289.
95. E. M. Golding, *Brain Res. Brain Res. Rev.*, 2002, **38**, 377.
96. D. I. Graham, J. H. Adams and E. Doyle, *J. Neurol. Sci.*, 1978, **39**, 213.
97. D. I. Graham, I. Ford, J. H. Adams, D. Doyle, G. M. Teasdale, A. E. Lawrence and D. R. McLellan, *J. Neurol. Neurosurg. Psychiatry*, 1989, **52**, 346.
98. E. M. Golding, C. S. Robertson and R. M. Bryan, Jr, *Clin. Exp. Hypertens.*, 1999, **21**, 299.
99. J. J. Bazarian, J. McClung, Y. T. Cheng, W. Flesher and S. M. Schneider, *Emerg. Med. J.*, 2005, **22**, 473.
100. A. S. Jagoda, J. J. Bazarian, J. J. Bruns, Jr, S. V. Cantrill, A. D. Gean, P. K. Howard, J. Ghajar, S. Riggio, D. W. Wright, R. L. Wears, A. Bakshy, P. Burgess, M. M. Wald and R. R. Whitson, *Ann. Emerg. Med.*, 2008, **52**, 714.
101. J. D. Lewine, J. T. Davis, E. D. Bigler, R. Thoma, D. Hill, M. Funke, J. H. Sloan, S. Hall and W. W. Orrison, *J. Head Trauma Rehabil.*, 2007, **22**, 141.
102. M. Inglese, R. I. Grossman, L. Diller, J. S. Babb, O. Gonen, J. M. Silver and H. Rusinek, *Brain Inj.*, 2006, **20**, 15.
103. R. A. Yeo, C. Gasparovic, F. Merideth, D. Ruhl, D. Doezema and A. R. Mayer, *J. Neurotrauma*, 2011, **28**, 1.
104. K. K. Holli, M. Wäljas, L. Harrison, S. Liimatainen, T. Luukkaala, P. Ryymin, H. Eskola, S. Soimakallio, J. Ohman and P. Dastidar, *Acad. Radiol.*, 2010, **17**, 1096.
105. K. K. Holli, L. Harrison, P. Dastidar, M. Wäljas, S. Liimatainen, T. Luukkaala, J. Ohman, S. Soimakallio and H. Eskola, *BMC Med. Imaging*, 2010, **10**, 8.

Immunoexcitotoxicity as a Central Mechanism of Chronic Traumatic Encephalopathy – A Unifying Hypothesis

RUSSELL L. BLAYLOCK, MD*[a] AND
JOSEPH C. MAROON, MD[b]

[a] Theoretical Neurosciences, LLC, Visiting Professor Biology, Belhaven University, Jackson, MS, USA; [b] Vice Chairman and Clinical Professor, Heindl Scholar in Neuroscience, Department of Neurosurgery, University of Pittsburgh Medical Center and Team Neurosurgeon, The Pittsburgh Steelers, Pittsburgh, PA, USA
*Email: blay6307@bellsouth.net

3.1 Introduction

Approximately 1.7 million people in the United States annually experience a traumatic brain injury (TBI).[1] The number of unreported head injuries is much higher. Sports-related head injuries account for 100 000 to 300 000 concussions in the game of football alone.[2] Most of these injuries are considered minor and a significant number are repeated injuries occurring over a relatively short period. It is known that professional contact sports participants experience thousands of sub-concussive blows during a career.[2,3]

Until recently, it was assumed that minor injuries resulted in few long-term neurological problems and were in fact characterized by a lack of

RSC Drug Discovery Series No. 24
Biomarkers for Traumatic Brain Injury
Edited by Svetlana A. Dambinova, Ronald L. Hayes and Kevin K. W. Wang
© Royal Society of Chemistry 2012
Published by the Royal Society of Chemistry, www.rsc.org

neuropathological changes in the brain. Recently, Omalu *et al.* described a set of pathological changes in professional football players that resembled at least some of the changes seen in the Alzheimer's brain.[4,5] He named this condition "chronic traumatic encephalopathy", or CTE. Similar pathological events are now known to occur in soldiers exposed to the compressive forces released from explosive devices.

Some individuals suffering from mild traumatic brain injuries, especially repetitive mild concussions, are thought to develop this slowly progressive encephalopathy, but not all. A central pathological mechanism explaining the development of progressive degeneration in this subset of individuals has not been elucidated. Yet, a large number of studies indicate that a process called immunoexcitotoxicity may be playing a central role in many neurodegenerative diseases, including chronic traumatic encephalopathy (CTE). The term "immunoexcitotoxicity" was first coined by the lead author to explain the evolving pathological and neurodevelopmental changes in autism and the Gulf War Syndrome, but it can be applied to a number of neurodegenerative disorders as well.[6]

The interaction between pro-inflammatory cytokines and glutamate type receptors trigger a series of neurodestructive events, such as extensive generation of reactive oxygen and nitrogen species, elevations in inflammatory PGE2, high levels of lipid peroxidation products, elevations in nitric oxide and superoxide (leading to peroxynitrite generation), and eventual neurodegeneration.[7–9] These events result in dendritic retraction, damage to axonal and dendritic microtubules, mitochondrial suppression, synaptic loss and eventually neuronal dropout.

While many studies have discussed the immune reactions and excitotoxicity associated with traumatic brain injury (TBI), only rarely are they discussed together or is a link made between the two processes.[10–18] The lead author first explored the effects of immune-induced excitotoxicity to explain many of the pathological changes being described with autism spectrum disorders and Gulf War Syndrome.[19,20] Excitotoxicity can indeed explain most of these changes and even the progressive nature of the repetitive injuries, but the story was still incomplete without an explanation of the immune findings and their relation to the damage.

It should be noted that the role of shearing forces is significantly more important in regards to immediate damage seen with moderate and severe head injuries than with minor injuries to the brain. This mainly includes shearing of selected white matter tracts. Detailed diffusion tensor imaging (DTI), suggests that even in minor head injuries, progressive axonal damage and eventual shearing is occurring.[21] As this chapter discusses, with repetitive minor TBIs, a series of progressive immunoexcitotoxic events occur, not only involving neuronal damage, but also delayed, progressive synaptic loss, dendritic retraction, and severing of axons as well.

Extensive neuropathological examinations of the CTE brain indicates that rather than the expected predominance of amyloid plaques, over half the cases have dense accumulations of hyperphosphorylated tau proteins in selected areas of the brain.[22] The tau proteins are identical to that seen in cases of Alzheimer's disease.[22] No one has offered an explanation for these unusual findings apart from references to oxidative stress and other less defined mechanisms, that in itself does not explain this particular pathological picture and especially its dynamic presentation. Our hypothesis explains most of the dynamic pathological and behavioral findings in CTE.

3.2 Clinical Features of Chronic Traumatic Encephalopathy

Martland in 1928 introduced the idea of a chronic progressive degeneration of the brain following repeated blows to the head in boxers, which he called the "punch drunk syndrome".[23] The findings of similar pathology in others exposed to contact sports by Omalu *et al.* suggested this was a more widespread problem.[5] McKee and others hypothesized that the pathological and clinical picture may be secondary to multiple sub-concussive blows to the head.[22]

The presentation of the syndrome is rather insidious in its earliest stages, usually presenting with problems with recent memory, poor attention, bouts of disorientation, depression, confusion, irritability, and frequent headaches.[22] The rate of progression varies, but usually there is a progressive deterioration of mental abilities. Poor insight and judgment, worsening of disorientation and confusion, and the onset of a number of antisocial behaviors are characteristic, including outburst of anger, suicidal mentation, apathy, insomnia, and irritability. A number of these symptoms are related to poor prefrontal cortex executive function and limbic dysfunction.

In severe, more advanced cases, one may see progressive slowing of motor movements, a staggering, propulsive gait, masked faces, impaired speech, tremors, vertigo, and deafness. A number of patients show advanced brainstem injury with prominent dysarthria, dysphagia, and ocular abnormalities. Clinically, as pathologically, one sees widespread involvement of the central nervous system, including the spinal cord.[22] A number of patients develop overt symptoms of Parkinsonism.[22]

McKee and others have noted that the pathology of CTE differs mainly in distribution, which one would expect with Alzheimer's disease beings a spontaneous disorder, whereas CTE is secondary to widespread brain injury.[22] With the football players, the playing time varied from 14 to 23 years. One third were symptomatic at the time of their retirement and half were symptomatic within 4 years after ending active play. In most cases there was a slow progressive decline in neurological function lasting decades.

3.3 Pathological Features of Chronic Traumatic Encephalopathy

McKee *et al.* examined 51 brains of CTE victims and found a number of neuropathological changes.[22] They noted marked atrophy of the entorhinal cortex, hippocampus, and amygdala in a number of the cases, which was accompanied by very dense gliosis and neurofibrillary degeneration. With advanced disease, they observed severe neuronal loss in the subcallosal and insular cortex, as well as mammillary bodies, medial thalamus, substantia nigra, locus ceruleus, and nucleus accumbens.

Neurofibrillary tangles (NFTs), astrocytic tangles, and dot-like spindle-shaped neuropil neurites (NNs), indicating severe neuronal injury and degenerating neurons, were commonly seen in the dorsolateral frontal, subcallosal, insular, temporal, dorsolateral parietal, and interior occipital corticies. There was a consistent appearance of perivascular tau-immunoreactive particles irregularly distributed in the cortex, primarily the superficial layers.

They also found involvement of the white matter, which was less severe than the cortex. NNs and fibrillar astrocytic tangles were found in the corpus collosum and subcortical white matter, especially the u-shaped fibers. Extremely dense NFTs with ghost tangles and severe neuronal loss were seen in the entorhinal cortex, hippocampus, and amygdala. The abnormal tau proteins, as stated, were pathologically indistinguishable from that seen with AD.

3.4   Microglial Priming and Activation with TBI

Microglial and astrocytic activation are commonly reported with TBI.[24–26] Experimental studies show that microglial activation occurs rapidly and is easily triggered, even with minor disturbances of brain homeostasis.[27] Microglial activation itself does not necessarily denote neurodestruction. It has been proposed that there are three states of activation: neurotropic/phagocytic that is neuroprotective, a predominately neurodestructive phenotype, and an intermediate phenotype in which both neurodestructive and reparative elements are released.[28] Regulation of these various states is still mostly unknown, but we do know that cytokine and glutamate levels play an important role.

Microglia, when fully activated, can release a number of reactive molecules, including pro- and anti-inflammatory cytokines, chemokines, interferons, prostaglandins, nitric oxide, proteases (metalloproteases), reactive oxygen and nitrogen species, and three excitatory neurotransmitters – glutamate, aspartate, and quinolinic acid (QUIN) (Figure 3.1).[29–31] In an inactive or ramified state, they possess long processes that exhibit continuous motility and secrete a number of neurotrophic factors, including brain-derived neurotropic factors (BDNF), neurotropic factor (NTF), and basic fibroblastic neurotropic factor (bFNF).[31]

Microglial membranes contain a number of receptors, such as cytokine, chemokine receptors, RANTES (regulated upon activation, normal T-cell expressed and secreted), receptor for advanced glycation end products (RAGE), NMDA receptors, AMPA/kainate receptors and metabotropic glutamate receptors, cholinergic receptors, P2 receptors, and opiate receptors, thus reacting to a great number of signals in specific ways.[30,31]

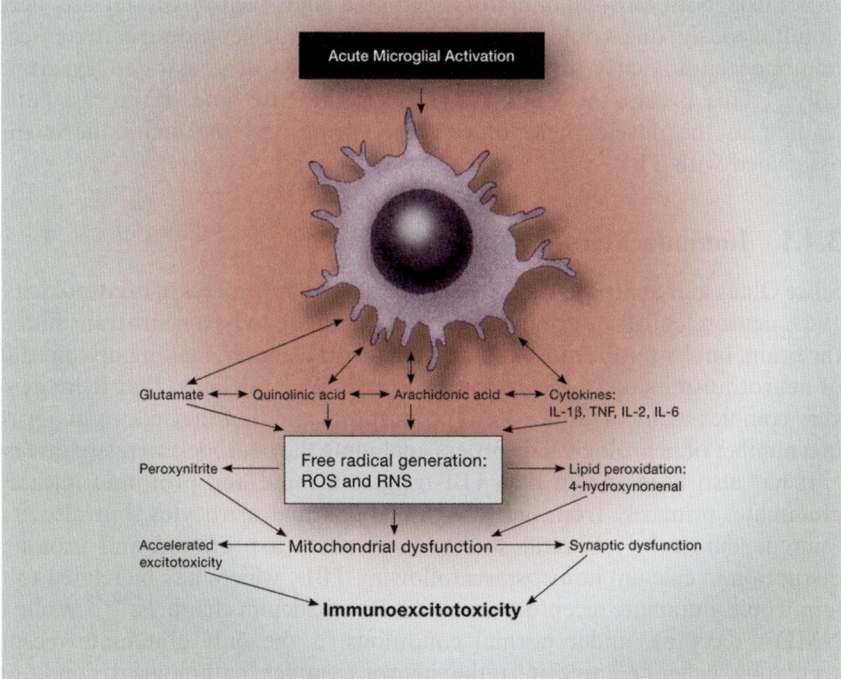

Figure 3.1 Illustration of the neurotoxic factors released from an activated microglia, demonstrating the interaction of proinflammatory cytokines and excitatory amino acids. Of particular importance is the effect on mitochondrial function, which when depressed enhances excitotoxic sensitivity as well as reactive oxygen species generation.

When in a neurodestructive phenotype, microglia co-release immune factors and excitotoxins. In addition, each can stimulate the others release. It is important to appreciate that elevations in individual pro-inflammatory cytokines is in itself not harmful to neurons, whereas in combination, such as IL-1ß and TNF-alpha, destructive reactions can be robust.[32] Growing evidence suggest that it is the combination of pro-inflammatory cytokines and excitotoxins that is producing the damage, rather than the cytokines alone. This is demonstrated by studies which show that subtoxic concentrations of LPS and subtoxic concentrations of glutamate, while not neurotoxic alone, are fully neurotoxic when combined.[33,34]

There is evidence that it is the excitotoxic component that is most neurodestructive. For example, Morimoto *et al.* found that a co-injection of LPS (immune stimulation) plus ibotenate, an NMDA receptor agonist, led to significant neuronal destruction and severe tissue collapse, but blocking excitotoxicity prevented any tissue damage, despite substantial microglial activation.[35] If ibotenate was given one day after the LPS injection, gross microglial activation occurred along with significant neurodegeneration. Other studies have also confirmed the primary role being played by excitotoxicity with immune stimulation.[36,37]

Because both pro-inflammatory cytokine and excitotoxin release occurs simultaneously, one would expect to see a synergistic neurodestructive cascade, immunoexcitotoxicity, set in motion, as has been demonstrated experimentally.[38] This process also adversely affects the BBB and brain vasculature, leading to the development of cerebral edema and the metabolic changes associated with TBI.[39,40]

3.4.1 Immunoexcitotoxicity

Since Olney and Sharpe first described and named the process of excitotoxicity in 1969, neuroscientists not only discovered glutamate to be a neurotransmitter in the brain, but by far the most abundant neurotransmitter, making up some 90% of neurotransmission in the cerebral cortex.[41] We now know that there exist a very complex array of glutamate-type receptors and that they play a major role in a number of neurological disorders, including the neurodegenerative diseases.

It has also been shown that TBI triggers a rapid and profound release of glutamate, primarily from activated microglia and astrocytes, but also from injured neurons.[42,43] Several studies have noted a profound and prolonged disruption in calcium homeostasis following TBIs, which may be related to the ionotropic glutamate receptors as regulators of calcium channels.[44,45] While the NMDA receptor, under normal conditions, is the only glutamate receptor regulating calcium entry into the neurons, under pathological conditions, GluR2-lacking AMPA receptors are trafficked to the synaptic membrane (Figure 3.2).[46] These special AMPA receptors do regulate calcium channels. Normally, AMPA receptors contain the GluR2 subunit and do not regulate the calcium channels. GluR2-lacking AMPA receptors have been seen in a number of neurological conditions.[47–51]

It is with the effect of immune cytokines on glutamate receptor physiology that one sees the interaction of pro-inflammatory immune cytokine receptors and excitototicity. Studies have shown that tumor necrosis factor-α (TNF-α) can enhance the trafficking of AMPA receptors to the synaptic membrane, which not only magnifies neurotransmission, but also worsens excitotoxicity under pathological conditions.[52–54] TNF-α operates through two receptors, TNFR1 and TNFR2. The former are mainly located on neuron membranes and the latter on glial membranes. TNFR1 activation triggers the trafficking of GluR-lacking AMPA receptors and is therefore potentially neurotoxic and TNFR2 is mainly neuroprotective.[55] Low levels of TNF-α activate TNFR2 so as to protect the glia and high levels activate the neurodestructive TNFR1.

Pro-inflammatory cytokines are also known to stimulate recruitment of microglia in the surrounding brain. Another connection between the immune cytokines and excitotoxicity is the observation that high levels of TNF-α up-regulate the enzyme glutaminase, which increases the production and release of glutamate via hemichannels.[56,57] The primary source of both immune cytokines and excitotoxins are the activated microglia, with significant contributions from astrocytes. The microglia are activated first and regulate astroglial activation and function.

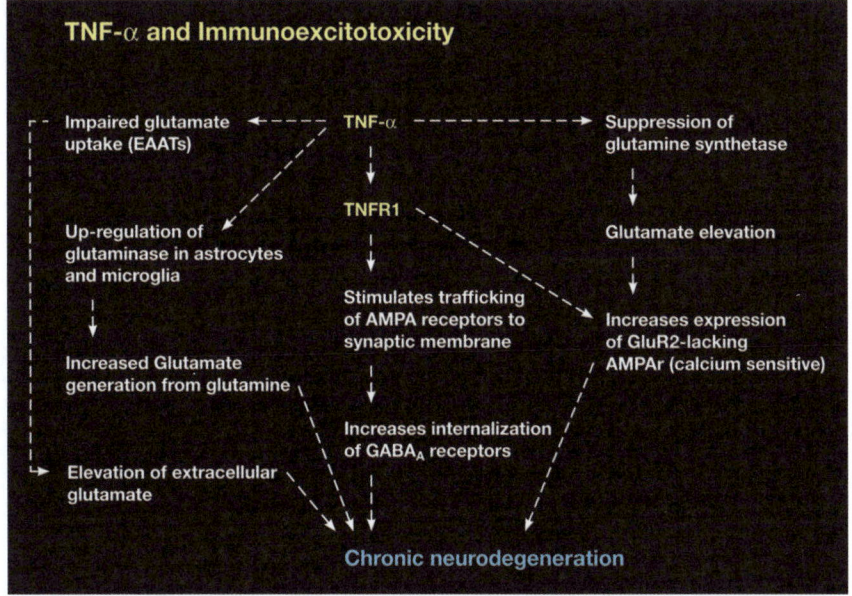

Figure 3.2 Diagram demonstrating a number of the major mechanisms of immunoexcitotoxicity, which includes the interaction of TNF-α with a number of systems that enhance excitotoxicity. This includes impaired glutamate transport, up-regulation of glutaminase, suppression of glutamine synthetase, increased trafficking of AMPA receptors to synaptic lipid raft, and endocytosis of GABA.

Disturbances of brain homeostasis triggers very rapid microglial activation from their resting or ramified state. Activated microglia can appear in a number of functional states, from neuroreparative or primarily neurodestructive, depending on a number of conditions. As the brain's primary innate immune cell, the microglia must not only activate rapidly, but it must also switch to a reparative mode and finally a ramified state. It appears that the principle triggers for this switching process to an inactive state include CD200 and fractalkines.[58–60] Several studies have shown that deficiencies in these switching regulators are associated with intense inflammatory injury to the brain (Figure 3.3).[61] CD200 receptors are deficient in cases of AD and PD and also decline with aging.[61,62]

Animal studies demonstrate that mice lacking CD200 receptors, for example, demonstrate exaggerated pro-inflammatory cytokine release with LPS stimulation. CD200 deficiencies have been reported with surgical trauma in aged rats.[63] Mascocha *et al.* demonstrated that mice given lipopolysaccharide (LPS) i.p., a potent immune stimulator, continued to have elevated microglial activation as long as 1 year after LPS exposure and that CD200 mRNA was decreased at 1 year.[64] They suggested that changes in CD200/CD200 receptor equilibrium might keep microglia chronically activated. Another study, found

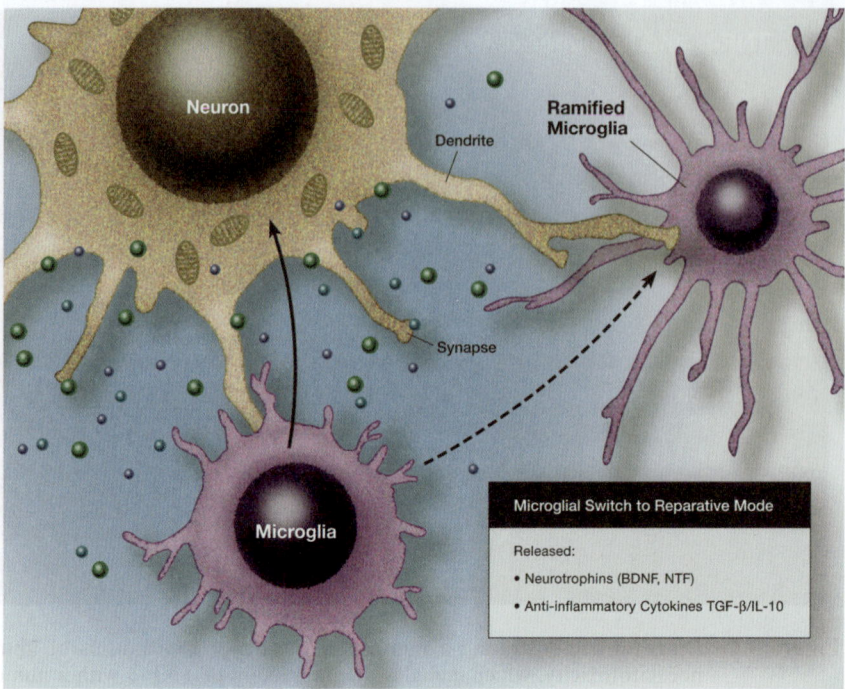

Figure 3.3   Illustration of an activated microglia that fails to switch from an activated, neurodestructive mode to a reparative mode or ramified state. Under such conditions immunoexcitotoxic reactions can continue for prolonged periods.

that wild type mice have attenuated disease as models of experimental allergic encephalomyelitis (EAE), with decreased CNS macrophage and microglial activation and reduced demyelination and axonal pathology. Interestingly, these mice show robust expression of CD200 on their neurons.[65]

With the initial injury, microglia are often switched to a primed state, in which they are geared up for intense activity, but release few cytokines or glutamate. With a second injury occurring with weeks to even years, these primed microglia release much higher levels of inflammatory cytokines than unprimed microglia and potentially do so for very long periods after the initiating insult.[66,67] This would explain the progressive nature of the degeneration associated with some cases of repeated minor head injuries. Should the person have a defect in the microglia switching mechanism, the microglia could remain in the neurodestructive mode and the damage would progress over a prolonged period. This effect has been seen with certain neurotoxins and neurological conditions.[67–69]

3.4.2   The Aging Brain and Immunoexcitotoxicity

Over time, as the brain ages, microglia become spontaneously primed, leading to a magnification of the degenerative process.[70] This would explain why the football players were seen to deteriorate cognitively years after ending play. It has also been shown that systemic immune activation worsens brain pathology, especially in the aging brain.[71,72] Exposure to neurotoxic environmental elements (pesticides/herbicides, neurotoxic metals, other head injuries, etc.) and recurrent infections or excessive immunizations could play a role in accelerating the damage.[71–75] Stress itself can also increase microglial priming and activation, as well as increase brain reactive oxygen species/reactive nitrogen species (ROS/RNS) and lipid peroxidation (LPO).[76,77] IL-1ß is the main activation factor for microglia, even though other elements can also activate them, including interferon-γ and caspase-3.[78,79] Elevations of systemic IL-1ß can produce rapid and widespread microglial priming and activation in the central nervous system (CNS). There are a number of pathways connecting the systemic immune system to the CNS, primarily the circumventricular organs (CVO) and vagus nerve.

The immunoexcitotoxic cascade eventually becomes a self-generating process, since it triggers high levels of brain oxidative stress, elevated lipid peroxidation products, mitochondrial suppression, and cell membrane damage, all of which interfere with glutamate clearance from the extraneuronal space. As one can see, this explains the calcium dysregulation described with even moderate head injuries, as well as the progressive neurodegeneration. For example, Sun and co-workers using a fluid concussion model for moderate TBI found that calcium remained elevated in CA3 hippocampal neurons for 30 days and never returned to baseline levels.[80] The calcium elevation was not secondary to cell death, but rather was shown to result from over-activation of NMDA and AMPA receptors by glutamate. Sun suggested that calcium dysregulation seen with moderate TBI might be permanent.

Another system, which can increase excitotoxic damage, is the cystine/glutamate Xc antiporter.[81] When extraneuronal glutamate levels are elevated, the exchange of intraneuronal glutamate for cystine is reduced. Cystine is essential for formation of brain glutathione. As neuronal glutathione levels fall, the affected neurons become infinitively more susceptible to damage by oxidative stress. Newer studies are showing that the cystine/glutamate antiporter is playing a major role in excitotoxic neurodegeneration.[82–84]

Aging is associated with a number of conditions that would accelerate neurodegeneration, such as mitochondrial dysfunction, a loss of membrane fluidity, microglial and astrocytic dystrophy, impaired autophagy, impaired antioxidant enzyme systems, endoplasmic reticulum stress, accumulation of DNA damage, impaired DNA repair, impaired blood-brain barrier function, vascular abnormalities, systemic immune aging with associated brain immune dysfunction, and progressive lowering of neuronal glutathione secondary to impaired Xc antiporter function.[66,85,86] Most of these processes are also connected to the immunoexcitotoxic process.

3.4.3 White Matter Damage with Immunoexcitotoxicity

One of the most commonly described pathological changes in the traumatized brain is diffuse axonal injury. More recent studies have shown that the injury to axons was not diffuse, but rather affected specific white matter tracts based on the severity of the injury.[87] Interestingly, it has been shown that with mild injury, most of the initial damage to the axons was not anatomical shearing, but rather involved a progressive degenerative process with any severing of axons occurring as a secondary event.[88] That is, most of the axonal damage was neurodegenerative and not mechanical.

There is compelling evidence that an immunoexcitotoxic mechanism is at play in this process, since one sees both an elevation in inflammatory cytokines and elevated levels of glutamate in the vicinity of the damage, as in the case of multiple sclerosis, spinal cord injury, and EAE, and is magnified in the aging CNS.[89–92] Axons are known to contain numerous glutamate receptors, as do oligodendrocytes, with AMPA/kainate receptors being the most abundant.[93,94] Microglia activation in the vicinity of axons has also been described.[89] In essence, a progressive shearing of axons has been described in multiple sclerosis secondary to an immunoexcitotoxicity process. Oligodendrocytes contain membrane glutamate receptors and undergo excitotoxic degeneration when extraneuronal levels are sufficiently elevated or sensitivity of the glutamate receptors is increased, as with the cytokine crosstalk previously described.

There is growing evidence that AD, autism, and other cognitive disorders involve a connectivity problem secondary to white matter injury. Recent technology, such as diffuse tenor imaging (DTI), allows researchers to view individual fiber tracts in the living brain. Recent DTI studies have indeed shown occult white matter injury in cases of moderate TBI, not seen with traditional magnetic resonance imaging (MRI) scanning techniques.[95] Rodriguez *et al.*, using a moderate parasagittal fluid-percussion injury model, found chronic axonal changes accumulating as late as 6 months after injury.[96] They examined the fimbria, external capsule, thalamus, and cerebral cortex, ipsilateral to the injury, using electron microscopy (EM) and light microscopy and found a progressive loss of axons over time with significant macrophage/ microglial infiltration as late as 6 months after injury.

3.5   Animals and Human Studies Showing Delayed, Progressive Axonal Injury

The question to be asked is – does repetitive brain injury trigger an immu- noexcitotoxic response that is prolonged and does this result in progressive states of neurodegeneration of axons? As discussed, a number of animal studies demonstrate a chronic degenerative process involving axons following TBI. Iwata and co-workers, using a swine TBI model, found only limited axonal Aß accumulation within axons appearing acutely after injury, but observed greater accumulation 1 month after injury that persisted for at least a year.[97] Chen *et al.*, also using a swine TBI model, found evidence of ongoing axonal pathology 6 months after rotational brain injuries.[98,99]

Human studies also demonstrate progressive white matter atrophy occurring after traumatic brain injuries.[95] Several studies have related this axonal injury to an accumulation of glutamate from activated microglia and astrocytes.[100] Studies have also shown that delayed axonal injury was worse in older animals, in keeping with previous observations of aging-related priming effects and immunoexcitotoxicity.[92] In older animals one sees a greater release of glutamate secondary to a reverse Na^+-dependent glutamate transport from GLT-1 glutamate transporters. With aging, there is a dysfunction of GLT-1 glutamate transporters.[101,102] Quardouz and co-workers demonstrated pathological intra-axonal elevations of calcium triggered by stimulating AMPA/kainate receptors on axons, which they concluded could result in progressive axonal degeneration with white matter injury.[103]

One would expect greater delayed and progressive axonal injury in cases of repetitive injury, based on the observation of prolonged microglial priming and impaired switching of microglia to the ramified phenotype seen with other conditions.

3.5.1 Macrophage Recruitment to the CNS

Release of chemokines, especially chemoattractant protein-1 (MCP-1), also known as CCL2, stimulates the recruitment of peripheral monocytes/macrophages to the CNS. Entry into the CNS occurs mainly via the CVO, especially the choroid plexus.[104] With more severe injury to the brain, other leucocytes are recruited as well. There is evidence that recruitment of regulatory T-lymphocytes (T_{regs}) to the CNS plays a major role in containing inflammation by interacting with activated microglia.[105,106] There is some evidence that recruited macrophages, which transform into brain microglia, play a more important role in brain neurodestruction than do resident microglia. Even peripheral macrophages secrete glutamate, as well as immune factors.[107] Recruited macrophages can spread throughout the brain parenchyma.

3.6 Immunoexcitotoxicity and Hyperphosphorylation of Tau

One of the mysteries of CTE is why are we seeing such dramatic accumulations of hyperphosphorylated tau in the damaged parts of the brain? It is known that excitotoxins can precipitate hyperphosphorylation of tau proteins (Figure 3.4).[108] This process has been demonstrated in studies using a variety of excitotoxins. One of the often-ignored excitotoxins is quinolinic acid (QUIN), a by-product of kynurenine breakdown during the process of tryptophan metabolism. Kynurenine is metabolized into kynurenic acid and picolinic acid, both of which are neuroprotective, and into QUIN, which is excitotoxic.

Activated microglia can secrete large amounts of QUIN, which acts on the NMDA receptor to trigger excitotoxicity when in excess.[109] Under conditions of brain inflammation, as we see in CTE, indolamine-2,3-deoxygenase (IDO) is up-regulated and this shifts kynurenine metabolism toward QUIN generation, a reaction which is especially driven by INF-γ.[110]

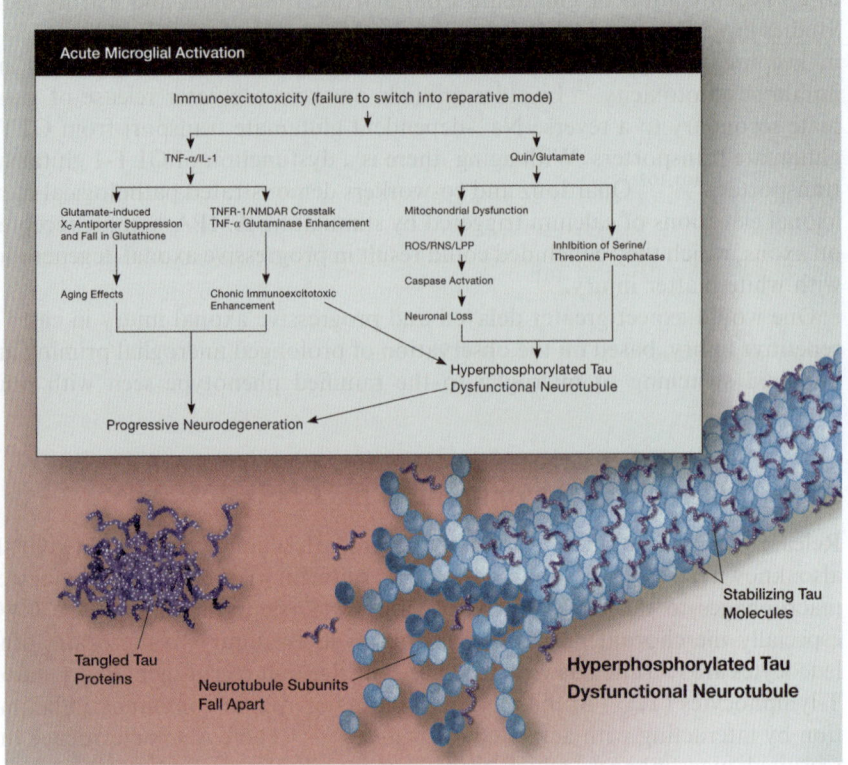

Figure 3.4   Illustration of the mechanism linking immunoexcitotoxicity with increased deposition of hyperphosphorylated tau in the CTE brain. Once the microglia fails to switch to the resting mode, elevated levels of IL-1ß and TNF-α act to enhance excitotoxicity. Neuronal glutathione levels fall secondary to inhibition of the glutamate/cystine antiporter X_c by excess glutamate, making the neuron significantly more sensitive to neurotoxic injury and death. High levels of glutamate suppress mitochondrial energy generation and this triggers necrotic and apoptotic cell death cascades. With chronic brain inflammation, tryptophan metabolism by the kynurenine pathway shifts toward quinolinic acid generation, which is excitotoxic and this results in hyperphosphorylated tau with subsequent neurotubule dysfunction and NFT deposition.

Elevations in QUIN can increase the generation of hyperphosphorylated tau, as we have seen with other excitotoxins.[111] Excitotoxins can also increase amyloid precursor protein (APP) processing, resulting in an increase in deposition of A beta in traumatized brain.[112] QUIN has also been shown to induce IL-1ß, a key cytokine in Alzheimer's disease and the prime activator of microglia.[113,114]

The kynurenine pathway is activated in the AD brain and other neurodegenerative diseases, leading to an accumulation of QUIN.[114] QUIN can be produced by microglia, astrocytes, and macrophages and immunostaining studies have shown high levels of QUIN reactivity in the perimeter of senile plaques and NFTs.[115] Both low and high levels of QUIN can cause structural changes in neurons.[116] For example, even concentrations as low as 100 nM in cultures can lead to dendritic beading, microtubule disruption, and a reduction in organelles.

In the AD brain, QUIN has been shown to accumulate in neurons, with high uptake in the entorhinal area and hippocampus.[116,117] It also co-localized with tau in cortical sections from AD patients. QUIN is found in significantly lower concentrations in the normal aging brain. *In vitro* treatment of human fetal neurons with submicromolar concentrations of QUIN significantly increased tau phosphorylation at multiple phosphorylation sites. Rahman *et al.* demonstrated that QUIN causes a decrease in expression of serine/threonine protein phosphatases and this leads to tau hyperphosphorylation.[111]

They also found that QUIN concentrations of 500 nM and 1200 nM significantly increased, not only total tau, but also tau phosphorylation at Tau-S and Tau-180 epitopes. QUIN increased phosphorylation of serine 199/202 and threonine 231 in a dose-dependent manner. All concentrations of QUIN tested decreased phosphatase activity by approximately 30% and were not dose dependent. QUIN also inhibits glutamate uptake and significantly inhibits glutamine synthetase activity in a dose-dependent manner, both of which increase excitotoxicity.

Glutamate and NMDA at equimolar concentrations (500 nM) also increased tau phosphorylation in a manner similar to QUIN.[108] Various NMDA receptor blockers have been shown to selectively inhibit tau phosphorylation (Memantine, MK-801, and AP-5).[108,118] This suggest that QUIN-induced tau hyperphosphorylation involves NMDA receptor activity.

Gorlovoy *et al.* found that co-culture of neurons with activated microglia induced the aggregation of tau neurites and that the treatment of neurons with TNF-α stimulated reactive oxygen species and accumulation of tau-citrine and tau-cerulean in neurites, which was inhibited by neutralization of TNF-α and the antioxidant trolox.[119] Liang and co-workers studied the dynamic changes in tau phosphorylation and tau-related protein kinase and protein phosphatase 2A (PP2A) in the mouse brain exposed to kainate (KA) excitotoxicity.[120] They found that KA led to transient dephosphorylation of tau protein (within 6 hours) followed by a sustained hyperphosphorylation of tau protein at multiple sites, sites also hyperphosphorylated in AD. They hypothesized that the initial dephosphorylation resulted from KA activation of PP2A and that the sustained hyperphosphorylated tau was mainly due to activity of cdk5 and down-regulation of PP2A.

Further analysis of this process had been done previously by Crespo-Biel and co-workers, who demonstrated, using subtoxic levels of KA, that KA cause a rapid and temporal phosphorylation (at 6 hours) of Erk and Akt dephosphorylation at 1 to 3 days following exposure.[121] Activation of GSK3beta increased on days 1–3 after KA exposure and was reduced by day 7. Caspase-3 also increased, but less so than calpain. Interestingly, calpain is involved in cdk5 activation and cdk5 is linked to GSK3beta activation. They demonstrated that tau hyperphosphorylation induced by excitotoxicity was due to an increase in GSK3beta/cdk5 activity in combination with inactivation of Akt. Subsequent activation of caspase-3 and calpain proteases led to dephosphorylation of tau, thus increasing microtubular destruction. These studies were done using a single exposure to excitotoxins, whereas in TBI there is a sustained elevation of brain extraneuronal glutamate/QUIN or hypersensitivity of synaptic and axonal glutamate receptors, which means that even physiological levels of glutamate can become excitotoxic, especially with mitochondrial impairment. This latter state is especially important. Many studies assume because glutamate levels are not elevated or moderately elevated, excitotoxicity is not a factor in producing the damage. Under a number of conditions, as we have seen, synapses can become hypersensitive to excitatory substances, such as hypoxia/ischemia, inflammation, mitochondrial dysfunction, and elevated levels of free radicals and lipid peroxidation products. In most instances, this hypersensitivity appears to involve an immunoexcitotoxic interaction between immune mediators and excitotoxins.

Hyperphosphorylation of tau has been shown to disrupt and disassemble microtubules and lead to neurodegeneration and memory loss. It is known that several neuroinflammatory mediators can activate the kynurenine pathway, leading to production of QUIN by activated microglia and invading macrophages. Microglial activation is also known to occur early in models of tauopathies and immune suppression attenuates tau pathology.[122] The lipid peroxidation product 4-hydroxynonenal (4-HNE) has been shown to prevent dephosphorylation of tau and promote tau cross-linking.[123] High levels of 4-HNE are generated by immunoexcitotoxicity and are associated with a number of neurodegenerative disorders.[124–127]

Taken together, there is convincing evidence that TBI, especially repetitive injuries, initiates the activation of innate brain immunity, which leads to transient immunoexcitotoxicity. Under normal conditions, this rapidly reverses as microglia assume a reparative phenotype. Preexisting brain pathology, even occult, or previous priming of microglia places the injured brain in a state of extended hyperreactivity, which can lead in some cases to a prolonged cascade of immunoexcitotoxic events, eventually culminating in progressive neurodegeneration.

As the brain ages, it becomes more vulnerable for a number of reasons, including progressive microglial activation and priming, attenuation of mitochondrial function, higher levels of inflammation and of longer duration, neuronal and glial dystrophy, reduced brain magnesium, impaired autophagy, periodic infections, and exposure to a number of neurotoxic environmental toxins and events.

Compelling evidence suggest that prolonged immunoexcitotoxicity can lead to hyperphosphorylation of tau, increased generation of neurotoxic levels of Aß oligomers, suppression of mitochondrial function and impaired migration to the synapse, increased generation of reactive oxygen and nitrogen species and lipid peroxidation products, all of which lead to synaptic and dendritic loss, neuronal apoptosis and necrosis, and progressive shearing of axons, pathological events commonly seen with CTE.

3.7 Conclusions

Rather compelling evidence, based on an enormous amount of basic research, suggest that immunoexcitotoxicity may represent a central mechanism explaining most of the pathological and clinical features of CTE. Previous studies confirmed a prolonged immunological reaction in the brain as well as excitotoxicity following TBI, which was more intense with repetitive injuries. Unfortunately, these two processes, for the most part, have been treated as separate, unrelated events. New research clearly demonstrates an interaction between pro-inflammatory cytokines and cytokine receptors and ionotropic glutamate receptor trafficking, primarily NMDA receptors and GluR2-lacking AMPA receptors.

Enhancement of excitotoxicity by pro-inflammatory process occurs on many levels other than receptor trafficking. For example, it has been shown that TNF-α can up-regulate the enzyme glutaminase in astrocytes, thus increasing the release of glutamate. Several studies have shown that inflammatory cytokines can also suppress EAATs and thus increase extraneuronal glutamate to excitotoxic levels. And by suppressing mitochondrial function, inflammatory cytokines also magnify the excitotoxic sensitivity of glutamate and QUIN. Oxidative glutamate toxicity must also be considered, which occurs when sustained elevations in extracellular glutamate impair the operation of the glutamate/cystine antiporter, leading to a substantial reduction in neuronal glutathione levels, with a much higher sensitivity to free radical and lipid peroxidation damage as well as sensitivity to the neurotoxic effects of certain metals.

The progressive nature of CTE following repetitive TBIs can be explained by the observation that prolonged activation of microglial in a neurodestructive phenotype does occur in certain pathological conditions. For example, pre-existing brain pathology, even occult, can be magnified with systemic immune stimulation. Also, it has been shown that with aging, brain microglia become spontaneously primed or activated and that under such conditions systemic immune stimulation is more intense and prolonged than at a younger age.

The presence of extensive hyperphosphorylated tau proteins in the injured areas of the brain can be explained by the observation that both glutamate and QUIN can cause hyperphosphorylation of tau, even at low concentrations, via suppression of phosphatases as well as other mechanisms. Both glutamate and QUIN are acting via the NMDA receptor to cause this effect. Because both excitotoxins generate 4-HNE, further interference with dephosphorylation of tau and an increase tau-crosslinking would be expected.

Only further research can confirm immunoexcitotoxicity as a central mechanism in CTE. A combination of dynamic DTI scanning and microglia activation imaging could go a long way in providing answers to some of these questions and demonstrating its dynamic features. If immunoexcitotoxicity is central to the process, then we may be able to take advantage of a number of pharmaceutical and natural products that have been shown to reduce this process, such as curcumin, doxycycline, minocycline, DHA, silymarin, and resveratrol.

References

1. X. L. Faul, M. M. Wald and V. Coronado, *Traumatic Brain Injury in the United States: Emergency Department Visits, Hospitalizations, and Deaths, 2002–2006,* CDC, National Center for Injury Prevention and Control, Atlanta, GA, 2010.
2. K. M. Guskiewicz, M. McCrea, S. W. Marshall, R. C. Cantu, C. Randolph and W. Barr, *JAMA*, 2003, **290**, 2549.
3. R. M. Greenwald, J. T. Gwin, J. J. Chu and J. J. Crisco, *Neurosurgery*, 2008, **62**, 789.
4. B. I. Omalu, S. T. DeKosky, R. L. Minster, M. I. Kamboh, R. L. Hamilton and C. H. Wecht, *Neurosurgery*, 2005, **57**, 128.
5. B. I. Omalu, S. T. DeKosky, R. L. Hamilton, R. L. Minster, M. I. Kamboh, A. M. Shakir and C. H. Wechtl, *Neurosurgery*, 2006, **59**, 1086.
6. R. L. Blaylock, *Alt. Ther. Health Med.*, 2009, **15**, 60.
7. J. H. Yi and A. S. Hazell, *Neurochem. Int.*, 2006, **48**(5), 394.
8. H. Slawik, B. Volk, B. Fiebich and M. Hull, *Neurochem. Int.*, 2004, **45**(5), 653.
9. P. E. Chabrier, C. Demerle-Pallardy and M. Auguet, *Cell Mol. Life Sci.*, 1999, **55**(8–9), 1029.
10. M. Rancan, V. I. Otto, V. H. Hans, I. Gerlach, R. Jork, O. Trentz, T. Kossmann and M. C. Morganti-Kossmann, *J. Neurosci. Res.*, 2001, **63**, 438.
11. M. C. Morganti-Kossmann, L. Satgunaseelan, N. Bye and T. Kossmann, *Injury*, 2007, **38**(12), 1392.
12. A. Czigner, A. Mihaly, O. Farkas, A. Buki, B. Krisztin-Peva, E. Dobo and P. Barzo, *Acta. Neurochir. (Wien)*, 2007, **149**(3), 281.
13. G. Chen, J. Shi, Z. Hu and C. Hang, *Mediators Inflamm.*, 2008, 716458.
14. K. Kamm, W. Vanderkolk, C. Lawrence, M. Jonker and A. T. Davis, *J. Trauma*, 2006, **60**(1), 152.
15. D. Bermpohl, Z. You, E. H. Lo, H. H. Kim and M. J. Whalen, *J. Cereb. Blood Flow Metab.*, 2007, **27**(11), 1806.
16. M. C. Morganti-Kossman, P. M. Lenzlinger, V. Hans, P. Stahel, E. Csuka, E. Ammann and R. Stocker, *Mol. Psychiatry*, 1997, **2**(2), 133.
17. A. M. Palmer, D. W. Marion, M. L. Botscheller, P. E. Swedlow, S. D. Styren and S. T. DeKosky, *J. Neurochem.*, 1993, **61**(6), 2015.
18. J. H. Yi and A. S. Hazell, *Neurochem. Int.*, 2006, **48**(5), 394.
19. R. L. Blaylock, *J. Am. Physc. Surg.*, 2004, **9**(2), 46.
20. R. L. Blaylock and A. Strunecka, *Curr. Med. Chem.*, 2009, **16**(2), 157.
21. M. Inglese, S. Makani, G. Johnson, B. A. Cohen, J. A. Silver, O. Gonen and R. I. Grossman, *J. Neurosurg.*, 2005, **103**, 298.
22. A. C. McKee, R. C. Cantu, C. J. Nowinski, E. T. Hedley-Whyte, B. E. Gavett, A. E. Budson, V. E. Santini, H. S. Lee, C. A. Kubilus and R. A. Stern, *J. Neuropathol. Exp. Neurol.*, 2009, **68**, 709.
23. H. S. Martland, *JAMA*, 1928, **91**, 1103.

24. M. Arand, H. Melzner, L. Kinzl, U. B. Bruckner and F. Gebhard, *Langenbecks Arch. Surg.*, 2001, **386**, 241.
25. B. M. Bellander, O. Bendel, G. Von Euler, M. Ohlsson and M. Vensson, *J. Neurotrauma*, 2004, **21**(5), 605.
26. S. Engel, H. Schluesener, M. Mittlebronn, K. Seid, D. Adjodah, H. D. Wehner and R. Meyermann, *Acta. Neuropathol.*, 2000, **100**(3), 313.
27. M. A. Lynch, *Mol. Neurobiol.*, 2009, **40**, 139.
28. G. C. Brown and J. J. Neher, *Biochem. Soc. Trans.*, 2007, **35**, 1119.
29. R. B. Banti, J. Gehrmann, P. Schubert and G. W. Kreutzberg, *Glia*, 1993, **7**, 111.
30. M. L. Block, L. Zecca and J. S. Hong, *Nat. Rev. Neurosci.*, 2007, **8**, 57.
31. R. M. Ransohoff and V. H. Perry, *Ann. Rev. Immunol.*, 2009, **27**, 119.
32. G. H. Jeohn, L. Y. Kong, B. Wilson, P. Hudson and J. S. Hong, *J. Neuroimmunol.*, 1998, **85**, 1.
33. H. M. Gao, J. S. Hong, W. Zhang and B. Liu, *J. Neurosci.*, 2003, **23**, 1228.
34. J. Hong, I. H. Cho, K. I. Kwak, E. C. Suh, J. Seo, H. J. Min, S. Y. Choi, C. H. Kim, S. H. Park, E. K. Jo, S. Lee, K. E. Lee and S. J. Lee, *J. Biol. Chem.*, 2010, **285**, 39447.
35. K. Morimoto, T. Murasugi and T. Oda, *Exp. Neurol.*, 2002, **177**, 95.
36. C. I. Rousset, J. Kassem, J. P. Oliver, S. Chalon, P. Gressens and E. Saliba, *J. Neuropathol. Exp. Neurol.*, 2008, **67**, 994.
37. A. Loddicks and N. J. Rothwell, *J. Cereb. Blood Flow Metab.*, 1996, **16**, 932.
38. I. Yawata, H. Takeuchi, Y. Doi, J. Lang, T. Mizuno and A. Suzumura, *Life Sci.*, 2008, **82**, 1111.
39. A. Gorgulu, T. Kitis, F. Unal, U. Turkoglu, M. Kucuk and S. Cobanoglu, *Acta Neurochir. (Wein)*, 1999, **141**, 93.
40. C. R. Kuhlmann, C. M. Zehendner, M. Gerigk, D. Closhen, B. Bender, P. Friedl and H. J. Luhmann, *Neurosci. Lett.*, 2009, **449**, 168.
41. J. W. Olney and L. G. Sharpe, *Science*, 1969, **166**, 386.
42. R. A. Ruppel, P. M. Kochanek, P. D. Adelson, M. E. Rose, S. R. Wisniewski, M. J. Bell, R. S. Clark, D. W. Marion and S. H. Graham, *J. Pediatr.*, 2001, **138**, 18.
43. J. H. Yi and A. S. Hazell, *Neurochem. Int.*, 2006, **48**, 394.
44. R. J. Hamm, D. M. O'Dell, B. R. Pike and B. G. Lyeth, *Brain Res. Cogn. Brain Res.*, 1993, **1**, 223.
45. L. S. Deshpande, D. A. Sun, S. Sombati, A. Baranova, M. S. Wilson, E. Attkisson, R. J. Hamm and R. J. DeLorenzo, *Neurosci. Lett.*, 2008, **441**(1), 115.
46. J. D. Bell, J. Ai, Y. Chen and A. J. Baker, *Brain*, 2007, **130**, 2528.
47. T. Opitz, S. Y. Grooms, M. V. Bennett and R. S. Zukin, *Proc. Natl. Acad. Sci. USA*, 2000, **97**(24), 13360.
48. S. Y. Grooms, T. Opitz, M. V. Bennett and R. S. Zukin, *Proc. Natl. Acad. Sci. USA*, 2000, **97**(7), 3631.
49. P. Van Damme, D. Braeken, G. Callewaert, W. Robberecht and L. Van Den Bosch, *J. Neuropathol. Exp. Neurol.*, 2005, **64**(7), 605.

50. B. Liu, M. Liao, J. G. Mielke, K. Ning and Y. Chen, L. Li, *et al.*, *J. Neurosci.*, 2006, **26**(20), 5309.
51. A. R. Ferguson, R. N. Christensen, J. C. Gensel, B. A. Miller, F. Sun, E. C. Beattie, J. C. Bresnahan and M. S. Beattie, *J. Neurosci.*, 2008, **28**(44), 11391.
52. A. M. Floden, S. Li and C. K. Combs, *J. Neurosci.*, 2005, **25**, 2566.
53. D. Leonoudakis, P. Zhao and E. C. Beattie, *J. Neurosci.*, 2008, **28**(9), 2119.
54. J. S. Park, N. Voitenko, R. S. Petralia, X. Guan, J. T. Xu, J. P. Sternberg, K. Takamiya, A. Sotnik, O. Kopach, R. L. Huganir and Y. X. Tao, *J. Neurosci.*, 2009, **29**(10), 3206.
55. D. Stellwagen, E. C. Beattie, J. Y. Seo and R. C. Malenka, *J. Neurosci.*, 2005, **25**, 3219.
56. T. F. Pais, C. Fegueiredo, R. Peixoto and M. H. Braz, *J. Neuroinflammation*, 2008, **5**, 43.
57. G. Takeuchi, S. Jin, J. Wang, G. Zhang, J. Kawanokuchi, R. Kuno, Y. Sonobe, T. Mizuno and A. Suzumura, *J. Biol. Chem.*, 2006, **281**(30), 21362.
58. L. C. Johnson, X. Su, K. Maguire-Zeiss, K. Horovitz, I. Ankoudinova, D. Guschin, P. Hadaczek, H. J. Federoff and K. Bankiewicz, *Mol. Ther.*, 2008, **16**(8), 1392.
59. A. Lyons, E. J. Downer, S. Crotty, Y. M. Noland, K. H. Mills and M. A. Lynch, *J. Neurosci.*, 2007, **27**, 8309.
60. A. Lyons, A. M. Lynch, E. J. Downer, R. Hanley, J. B. O'Sullivan, A. Smith and M. A. Lynch, *J. Neurochem.*, 2009, **110**(5), 1547.
61. D. G. Walker, J. E. Dalsing-Hernandez, N. A. Campbell and L. F. Lue, *Exp. Neurol.*, 2009, **215**(1), 5.
62. X. J. Wang, M. Ye, Y. H. Zhang and S. D. Chen, *J. Neuroimmune Pharmacol.*, 2007, **2**(3), 259.
63. X. Z. Cao, H. Ma, J. K. Wang, F. Liu, B. Y. Wu, A. Y. Tian, L. L. Wang and W. F. Tan, *Prog. Neuropsychopharmacology Biol. Psychiatry*, 2010, **34**, 1426.
64. W. Masocha, *J. Neuroimmunol.*, 2009, **214**(1–2), 78.
65. T. Chitnis, J. Imitola, Y. Wang, W. Elyaman, P. Chawla, M. Sharuk, K. Raddassi, R. T. Bronson and S. J. Khouri, *Am. J. Pathol.*, 2007, **170**, 1695.
66. R. N. Dilger and R. W. Johnson, *J. Leukoc. Biol.*, 2008, **84**, 932.
67. J. W. Langston, L. S. Fomo, J. Tetrud, A. G. Reeves, J. A. Kaplan and D. Karluk, *Ann. Neurol.*, 1999, **46**, 598.
68. E. N. Mango and S. Heyley, *Neurobiol. Aging*, 2009, **3**, 1361.
69. Y. Sekine, Y. Ouchi, G. Sugihara, N. Takei, E. Yoshikawa, K. Nakamura, Y. Iwata, K. J. Tsuchiya, S. Suda, K. Suzuki, M. Kawai, K. Takebayashi, S. Yamamoto, H. Matsuzaki, T. Ueki, N. Mori, M. S. Gold and J. L. Cadet, *J. Neurosci.*, 2008, **28**, 5756.
70. R. Sandhir, G. Onyszchunk and N. E. Berman, *Exp. Neurol.*, 2008, **13**, 372.

71. M. I. Combrinck, V. H. Perry and C. Cunningham, *Neuroscience*, 2002, **112**, 112.
72. R. Dantzer, *Nat. Rev. Neurosci.*, 2008, **9**, 46.
73. J. Y. Chang and L. Z. Liu, *Brain Res. Mol. Brain Res.*, 1999, **68**(1–2), 22.
74. T. B. Sherer, R. Betarbet, J. H. Kim and J. T. Greenamyre, *Neurosci. Lett.*, 2003, **341**(2), 87.
75. X. F. Wu, M. L. Block, W. Zhang, L. Qin, B. Wilson, W. Q. Zhang, B. Veronesi and J. S. Hong, *Antioxid. Redox Signal.*, 2005, **7**(5–6), 654.
76. H. Anisman, *Rev. Psychiatry Neurosci.*, 2009, **34**, 4.
77. L. Brydon, C. Walker, A. Wawrzniak, D. Whitehead, H. Okamura, J. Yajima, A. Tsuda and A. Steptoe, *Brain Behav. Immun.*, 2009, **23**, 217.
78. A. Basu, J. K. Krady and S. W. Levison, *J. Neurosci. Res.*, 2004, **78**(2), 151.
79. M. Masumura, R. Hata, I. Nishimura, T. Uetsuki, T. Sawada and K. Yoshikawa, *Brain Res. Mol. Brain Res.*, 2000, **80**(2), 219.
80. D. A. Sun, L. S. Deshpande, S. Sombati, A. Baranova, M. S. Wilson, R. J. Hamm and R. J. DeLorenzo, *Eur. J. Neurosci.*, 2008, **27**, 1659.
81. M. Domercq, M. V. Sanchez-Gomez, C. Sherwin, E. Etxebarria, R. Fern and C. Matute, *J. Immunol.*, 2007, **178**(10), 6549.
82. S. Qin, C. Colin, I. Hinners, A. Gervais, C. Cheret and M. Mallat, *J. Neurosci.*, 2006, **26**(12), 3345.
83. P. Albrecht, J. Lewerenz, S. Dittmer, R. Noack, P. Maher and A. Methner, *CNS Neurol. Disord. Drug Targets*, 2010, **9**(3), 373.
84. K. Mawatari, Y. Yasui, K. Sugitani, T. Takadera and S. Kato, *Neuroscience*, 1996, **73**(1), 201.
85. C. Gemma and P. C. Bickford, *Rev. Neurosci.*, 2007, **18**(2), 137.
86. G. Paradies, G. Petrosillo and F. M. Ruggiero, *Neurochem. Int.*, 2011, **58**(4), 447.
87. J. M. Melchior, J. D. Peduzzi, E. Elefheriou and T. A. Noverrack, *Arch. Phys. Med. Rehabil.*, 2001, **82**, 1461.
88. A. Buki and J. T. Povlshock, *Acta Neurochir. (Wein)*, 2006, **148**, 181.
89. O. W. Howell, J. L. Rundle, A. Garg, M. Komada, P. J. Brophy and R. Reynolds, *J. Neuropathol. Exp. Neurol.*, 2010, **69**(10), 1017.
90. E. Park, A. A. Velumian and M. G. Fehlings, *J. Neurotrauma*, 2004, **21**(6), 754.
91. J. H. Fowler, E. McCracken, D. Dewar and J. McCulloch, *Brain Res.*, 2003, **991**(1–2), 104.
92. J. E. Simpson, P. G. Ince, C. E. Higham, C. H. Gelsthorpe, M. S. Fernando, F. Matthews, F. Matthews, G. Forster, J. T. O'Brien, R. Barber, R. N. Kalaria, C. Brayne, P. J. Shaw, K. Stoeber, G. H. Williams, C. E. Lewis and S. B. Wharton, *Neuropathol. Appl. Neurobiol.*, 2007, **33**(6), 670.
93. P. K. Stys, *Curr. Mol. Med.*, 2004, **4**, 113.
94. S. B. Tekkok and M. P. Goldberg, *J. Neurosci.*, 2001, **21**, 4237.
95. R. Kumar, M. Husain, R. K. Gupta, K. M. Hasan, M. Haris, A. K. Agarwal, C. M. Pandey and P. A. Narayana, *J. Neurotrauma*, 2009, **26**, 481.

96. A. C. Rodrigues-Paez, J. P. Brunschwig and H. M. Bramlett, *Acta Neuropathol.*, 2005, **109**, 603.
97. A. Iwata, X. H. Chen, T. K. McIntosh, K. D. Browne and D. H. Smith, *J. Neuropath. Exp. Neurol.*, 2002, **61**, 1056.
98. X. H. Chen, V. E. Johnson, K. Uryu, J. Q. Trojanowski and D. H. Smith, *Brain Pathol.*, 2009, **19**, 214.
99. X. H. Chen, R. Siman, A. Iwata, D. F. Meaney, J. Q. Trojanowski and D. H. Smith, *Am. J. Pathol.*, 2004, **165**, 357.
100. D. P. Stirling and P. K. Stys, *Trend. Mol. Med.*, 2010, **16**(4), 160.
101. B. Potier, J. M. Billard, S. Rivere, P. M. Sinet, I. Denis, G. Champeil-Potokar, B. Grintal, A. Jouvenceau, M. Kollen and P. Dutar, *Aging Cell*, 2010, **9**(5), 722.
102. J. H. Yi, D. V. Pow and A. S. Hazell, *Glia*, 2005, **49**(1), 121.
103. M. Ouardouz, E. Coderre, A. Basak, A. Chen, G. W. Zamponi, S. Hameed, R. Rehak, X. Yin, B. D. Trapp and P. K. Stys, *Ann. Neurol.*, 2009, **65**, 151.
104. L. M. Maness, A. J. Kasten and W. A. Banks, *Am. J. Physiol.*, 1998, **275**, E207.
105. S. C. Byram, M. J. Carlson, C. A. Deboy, C. J. Serpe, V. M. Sanders and K. J. Jones, *J. Neurosci.*, 2004, **24**, 4333.
106. B. Melcor, S. S. Puntambeker and M. J. Carson, *Neurochem. Int.*, 2006, **49**, 145.
107. J. A. J. Hendricks, C. E. Teunissen, H. E. de Vries and C. D. Dijkstra, *Brain Res. Rev.*, 2005, **48**, 185.
108. Z. Liang, F. Liu, K. Iqbal, I. Grundke-Iqbal and C. X. Gong, *J. Alzheimer's Dis.*, 2009, **17**, 531.
109. M. P. Heyes, C. L. Achim, C. A. Wiley, E. O. Major, K. Saito and S. P. Markey, *Biochem. J.*, 1996, **320**, 595.
110. L. A. Pemberton, S. J. Kerr, G. Smythe and B. J. Brew, *Cytokine Res.*, 1997, **17**, 589.
111. A. Rahman, K. Ting, K. M. Cullen, B. J. Brew and G. J. Guillemin, *PloS One*, 2009, **4**, e6344.
112. N. W. Hu, I. Klyubin, R. Anway and M. J. Rowan, *Proc. Natl. Acad. Sci. USA*, 2009, **106**, 20504.
113. O. V. Forlenza, B. S. Diniz, L. L. Talib, V. A. Mendonca, E. B. Ojopi, W. F. Gattaz and A. L. Teixeira, *Demen. Geriatr. Cogn. Discord.*, 2009, **28**(6), 507.
114. G. J. Guillemin, K. R. Williams, D. G. Smith, J. Croitoru-Lamoury and B. J. Brew, *Adv. Exp. Med. Biol.*, 2003, **527**, 167.
115. G. J. Guillemin and B. J. Brew, *Redox Rep.*, 2002, **7**, 199.
116. S. J. Kerr, P. J. Arnati and B. J. Brew, *J. Neurovirol.*, 1995, **1**, 375.
117. G. J. Guillemin, B. J. Brew, C. E. Noonan, O. Takikawa and K. M. Cullen, *Neuropathol. Appl. Neurobiol.*, 2005, **31**, 395.
118. J. L. Moinuevo, A. Liado and L. Rami, *Am. J. Alzheimer's Dis. Other Demen.*, 2005, **20**, 77.
119. P. Gorlovoy, S. Larionov, T. T. Pham and H. Neumann, *FASEB J.*, 2009, **23**(8), 2502.

120. Z. Liang, F. Liu, K. Iqbal, I. Grundke-Iqbal and C. X. Gong, *J. Alzheimer's Dis.*, 2009, **17**(3), 531.
121. N. Crespo-Biel, A. M. Camins and M. Pallas, *Neurochem Int.*, 2007, **50**(2), 435.
122. Y. Yoshiyama, M. Higuchi, B. Zhang, S. M. Huang, N. Iwata, T. C. Saido, J. Maeda, T. Suhara, J. Q. Trojanowski and V. M. Lee, *Neuron*, 2007, **53**, 337.
123. M. P. Mattson, W. Fu, G. Waeg and K. Uchida, *Neuroreport*, 1997, **8**, 2275.
124. S. J. Siegel, J. Bieschke, E. T. Powers and J. W. Kelly, *Biochemistry*, 2007, **46**(6), 1503.
125. M. A. Lovell, W. D. Ehmann, M. P. Mattson and W. R. Markesbery, *Neurobiol. Aging*, 1997, **18**(5), 457.
126. L. Vigh, R. G. Smith, J. Soos, J. I. Engelhardt, S. H. Appel and L. Siklos, *Acta Neuropathol.*, 2005, **109**(6), 567.
127. K. Zarkovic, *Mol. Aspect. Med.*, 2003, **24**(4–5), 293.

CHAPTER 4

Neurodegradomics: The Source of Biomarkers for Mild Traumatic Brain Injury

SVETLANA A. DAMBINOVA

Kennesaw State University, WellStar College of Health & Human Services, 1000 Chastain Road, Kennesaw, GA 30144, USA
Email: sdambino@kennesaw.edu

4.1   Introduction

Biomarkers are defined as chemical substances that reflect the presence of a particular biological or medical condition.[1] In many clinical situations, additional knowledge regarding subclinical events in brain tissue may greatly assist diagnosis and treatment. Biomarkers may be used to monitor the time course of brain damage by detecting subclinical neuronal breakdown using blood assays.

One significant concern is uncertainty in diagnosing mild traumatic brain injury (TBI). Mild TBI impairs electrical and chemical circuits, subsequently leading to a minor failure in blood circulation or microvessel bleeding within CNS tissue. Minor biomechanical injury causes energy failure and protein synthesis/degradation processes in the brain, evoking a metabolic "storm" – namely, neurotoxicity – and, depending on the location of structural injury, can cause development of diverse CNS disorders.[2]

A key component in the control of the neurotoxicity cascade is a family of ionotropic glutamate receptors that consists of at least 25 membrane proteins

RSC Drug Discovery Series No. 24
Biomarkers for Traumatic Brain Injury
Edited by Svetlana A. Dambinova, Ronald L. Hayes and Kevin K. W. Wang
© Royal Society of Chemistry 2012
Published by the Royal Society of Chemistry, www.rsc.org

that direct signal transduction and neuronal communications.[3,4] In this chapter, results of the search for biomarkers of neurotoxicity to assess brain injury using molecular biological, biochemical, and immunochemical methods are described. How, when, and which specific glutamate receptor biomarker is released from the CNS into peripheral fluids in response to mild injury is explained using a neurodegradomics approach.

4.2 Neurodegradomics: The Source of Brain Biomarkers

Neurodegradomics, as a component of degradomics,[5] refers to the proteolytic breakdown of brain proteins to peptide fragments that can be detected in biological fluids. The search for specific biomarkers includes brain-specific proteases and protease-substrate products. The latter are integral substances for normal functions of nervous tissue and may be activated by pathological processes, causing abnormal protein degradation associated with certain neurological deficits. An understanding of the chemistry of major structural components, coupled with proteolysis in nervous tissue, helps in comprehending neurodegradomics.

Brain injury involves multi-systemic mechanisms of neurotoxic coupling, microvascular dysfunction, and immune responses that result in primary (acute) and secondary (chronic) events. CNS and systemic enzyme families, including serine (extracellular) and cysteine proteases (intracellular), and matrix metalloproteinases (blood-borne), are activated. Key metabolites of proteolytic cascades can be classified according to pathological conditions; i.e. ischemic, hemorrhagic, or axonal injury associated with abnormal spiking activity.

Currently, neurodegradomics – in contrast to degradomics – focuses primarily on studying the "fate" of cleaved products, rather than enzyme structure and functions. The major CNS substructures involved in neurodegradomic cascade are the blood-brain barrier (BBB), brain microvessels, and microglia. It appears that only proteolytic products comprising immunoactive epitopes released into the bloodstream are considered potential biomarkers of CNS pathology.

4.2.1 Blood-Brain Barrier

One important aspect in the search for brain biomarkers is the concept of BBB permeability in normal and pathological conditions. The brain is protected by endothelial and astroglial cellular barriers containing glutamate-induced metabolic pathways (enzymes, transporters, receptors, and mitochondrial energy activities). The tight junctions between the vascular endothelial cells and the relatively low vascular permeability effectively exclude most macromolecules from being delivered into or out of the brain. These junctions form the BBB that prevents brain chemicals, such as neurotransmitters, from leaking into the blood.[6] However, under normal conditions easy exchange of molecules, such as hormones, is facilitated by passage through the porous endothelium between the brain and the blood.

CNS biomarkers generally represent proteins that are specifically produced in the brain and spinal cord and are relatively sequestered in the CNS, mainly due to the presence of an intact BBB.[7,8] When the BBB is compromised due to brain injury or disorders, peptides and proteins are released into the biological fluids and may accumulate in the blood.[9,10] Therefore, detection and quantification of CNS-specific proteins released into the blood as a result of brain injury provides a potentially attractive means of diagnosing minor brain injuries using minimally invasive procedures.

4.2.2   Brain Microvessels

Another significant aspect of BBB functioning includes autoregulation of cerebral blood circulation that involves several types of traditional peripheral innervations, including sympathetic, parasympathetic, and sensory nerve fibers.[11] In addition, cerebral resistance may be innervated via glutamatergic pathways and receptors that exist entirely within the CNS. Cerebral blood vessels are exposed to major excitatory neurotransmitters (glutamate, aspartate) that escape into the extracellular fluid from synapses.[6] Sympathetic nerves appear to limit cerebrovascular dilation during arterial hypertension, hypoxia, and hypercapnia; attenuate disruption of the BBB during arterial hypertension; and contribute to the development and maintenance of vascular pathology.[12] Glutamate receptors have also been found to be located on the extracellular endothelial surface of brain microvessels, controlling their functions.[13] These receptors innervate adrenergic, cholinergic, and dopaminergic neural control systems, such as the locus ceruleus[14] and basal nucleus of Mynert.[15]

Incomplete denervation of axons, or temporary loss of CNS supply due to mild TBI, may cause short-term changes in blood flow circulation rate, blood pressure, and body temperature; these subtle, or asymptomatic, changes depend on the type of peripheral and central regulation.[16] Cerebrovascular reactions to mild TBI may be minor compared with those occurring with moderate or severe injuries, but even minor injuries can result in alterations in autoregulation of the vasculature and this can lead to ischemia/hypoxia.[17] It has been shown that a progressive loss of axonal integrity with chronic changes can accumulate as late as 6 months following TBI.[18] Several studies have related this axonal injury to an accumulation of glutamate from activated microglia and astrocytes.[19–21]

Analogous to muscle denervation causing over activation of acetylcholine receptors, leading to muscle atrophy, microvascular denervation results in up-regulation of glutamate receptors and consequent degradation of microvessel innervation. Depending on location of impairment, this degradation may cause either grey or white matter atrophy.[22] A number of chronic neurological conditions; i.e. neurodegenerative diseases, are associated with different brain and spinal cord structural atrophies that activate immune response.[23]

4.2.3 Brain Microglial System

The major resident immune cells within the CNS are the microglia, which protect nervous cells from glutamate-induced neurotoxicity by producing transporters that restore glutamate homeostasis.[24] When activated, microglia secrete a number of anti-inflammatory and pro-inflammatory cytokines, chemokines, nitric oxide, and prostaglandins and three excitotoxic molecules, glutamate, aspartate, and quinolinic acid.[25] Additionally, convincing evidence has shown that brain damage causes rapid microglial activation and/or proliferation, contributing to brain self-protection when activation is not prolonged or intense.[26] Secondary nervous tissue damage related to irreversible neurotoxic cascade can be correlated with high levels of oxidative stress, lipid peroxidation product accumulation, mitochondrial dysfunction, loss of calcium homeostasis, and proteolytic breakdown of CNS proteins, all of which are associated with neurological deficits.

4.2.4 Brain Proteolysis

There is growing evidence that proteolytic breakdown is a highly dynamic process involving brain proteases located extracellularly, intracellularly, and subcellularly.[27,28] Corresponding with a specific body organ, brain proteases can be either systemic (blood, cerebrospinal fluid, saliva) or organic (liver, brain, digestive system). Once activated, proteases initiate irreversible protein cleavage within the damaged extracellular or intracellular space. Within the CNS, proteases produce excessive amounts of multiple endogenous peptides in neuron-neuronal and/or neuron-glial contacts.[22]

The numerous proteases that regulate protein metabolism and degradation within the CNS are presented in Table 4.1. Concurrent analyses of proteolytic expression and dynamic changes in enzyme stimulation have demonstrated their role in a balance between activation and inhibition. When impaired, this balance is shifted to proteolytic activation with selective time-dependent processing from extrinsic to intrinsic enzyme reaction.[29] Intracellularly located cytoplasmic enzymes target erroneous and misfolding proteins in congruent locations,[30] whereas compartmental enzymes of subcellular structures are directed to cleavage of C-terminal domains of membrane proteins.[31] Numerous proteases control mechanisms of cell death after cerebral injury and produce a number of enzyme-dependent digestive neuropeptides that enter the bloodstream through the compromised BBB.

Table 4.1 CNS proteases.

Degradome	Nature and functions	Localization	Key process	References
Serine proteases	Cleave extracellular N-terminal	Glia, leukocytes	Neuronal degradation	*J. Biol. Chem.*, 2009, **284**, 12862
Cysteine proteases				
Calpains	Cytosolic endopeptidases, cleave to moderate substrate activity	Neurons, glial cells, and muscles	Generate peptide fragments with altered activity	*Sci. Signal.*, 2008, **1**, re1
Caspases	Cleave receptor intracellular C-terminals	Neurons, astrocytes, oligodendrocytes	Secondary injury an apoptotic cell death	*Arch. Neurol.*, 2000, **57**, 1273
Cathepsins		Neuronal and glial organelles	Apoptotic and necrotic cell death	*Curr. Pharm. Des.*, 2002, **8**, 1639
Matrix metalloproteinases (MMPs)	Family of zinc-endopeptidases, degrading extracellular matrix proteins	Neurons, glia, endothelial cells	Inflammation; may be active <3 hours of brain injury	*Curr. Opin. Cell. Biol.*, 2009, **21**, 645

4.3 Glutamate-Induced Neurotoxicity (Excitotoxicity) Triggers Secondary Brain Injury

Neurological complications after brain injury are initiated by metabolic disturbances, one of which is glutamate-induced neurotoxicity (excitotoxicity) (see Chapter 3). The term excitotoxicity is related to increased release and impaired uptake of excitatory amino acids, such as glutamate, aspartate, homocysteine, and quinolinic acid. The latter mediates toxic buildup of amino acids that leads to over-stimulation of glutamate receptors, which causes neuronal injury.

The excitotoxicity phenomena was discovered by Olney;[32] subsequent studies have disclosed a complex array of glutamate receptors, key components of excitotoxicity cascade.[4,33] This has triggered numerous studies of glutamate receptor pathophysiology to improve diagnostic, treatment, and prognostic strategies for neurological disorders.[10,22,34]

4.3.1 Biomarkers of Glutamate-Induced Neurotoxicity

Glutamate receptors (GluR) mediate a great majority of fast excitatory synaptic neurotransmission in the mammalian brain and their functions are absolutely critical for human higher brain functions including thinking, consciousness, memory, and learning. The most important is that these neuroreceptors regulate long- and short-term synaptic plasticity through adaptation via alterations in gene expression.[35] The latter is a major component of human behavioral adaptation to different environmental conditions.

GluR are a major family of approximately 25 proteins comprising channel-activated ionotropic or G-protein regulated metabotropic receptors, both of which initiate secondary messengers and are located on microvascular surfaces.[36] They also mediate neuron-glial contacts and glutamatergic neurotransmission[37] (see Table 4.2). More than 80% of cortical, subcortical, and spinal cord neuronal communications are dependent on structure, functions, and location of GluR.[4] There are three major ionotropic GluR subtypes: AMPA,[38] NMDA,[39] and kainite receptors,[40] which are categorized by structural similarity, electrophysiological functioning, and selective pharmacological properties (see Table 4.2). NMDA receptors mainly control Ca^{2+} channels, while non-NMDA (AMPA and kainite) receptors regulate permeability of all three ion channels (K^+, Na^+, and Ca^{2+}). Molecular diversity of GluR genes encoding various receptor subunits is responsible for pharmacological and functional heterogeneity. A number of reviews are available on the role of GluR CNS and their implication on the pathology of neuronal injury,[41] neuropsychiatric disorders,[42] and addiction.[43]

Recently, the prime response of plasmin to neurotoxicity through cleavage of the native N-terminal domain of NMDA receptor (NR2A subunit) was shown in an independent study.[44] Numerous layers of digestive endogenous products may be found in the brain and peripheral biological fluids.[45] For example, as shown in Figure 4.1, N-terminal domain AMPA receptors (112 kD) and several digestive products were revealed in brain and blood:[22] small peptides (5–7 kD),

Table 4.2   Ionotropic glutamate receptor family.

	Old/novel nomenclature		
Parameter	*NMDAR/GRIN 1-4*	*AMPAR/GRIA 1-4*	*Kainate Receptors/ GRIK 1-5*
Structure	NR1, NR2A, NR2B, NR3, NR4	GluR1, GluR2, GluR3, GluR4	GluR5, GluR6, KA1/KA2
Functions	Post-synaptic depolarization, STP, microvessel regulation	Fast post-synaptic EPSP, LTP, spiking activity	Pre-/post-synaptic miniature IPSP GABA transmission modulation
Major Distribution & Localization	Cortical: Forebrain, visual, motor, verbal areas, microvessel surfaces Cellular: extra- and intra-synaptic	Sub-cortical: hippo-campus, cerebellum, brain steam Cellular: dendrites and axons	Spinal cord, basal ganglia nucleus, dorsal root ganglions Cellular: sensor neurons, membranes
Disease related	Cerebral ischemia, glaucoma, dementia, schizophrenia, Alzheimer's, cocaine addiction	TBI, hypoxia, hypoglycemia, lead encephalopathy, epilepsy, heroin addiction	SCI, neuropathic pain, MS, Parkin-son', Huntington diseases, ALS, homocysteinuria
Pharmacologi-cal Trends	Memantine, anti-psychotics, neuropathic pain (antagonists)	AMPA-kines (modulators)	Anti-perceptive drugs

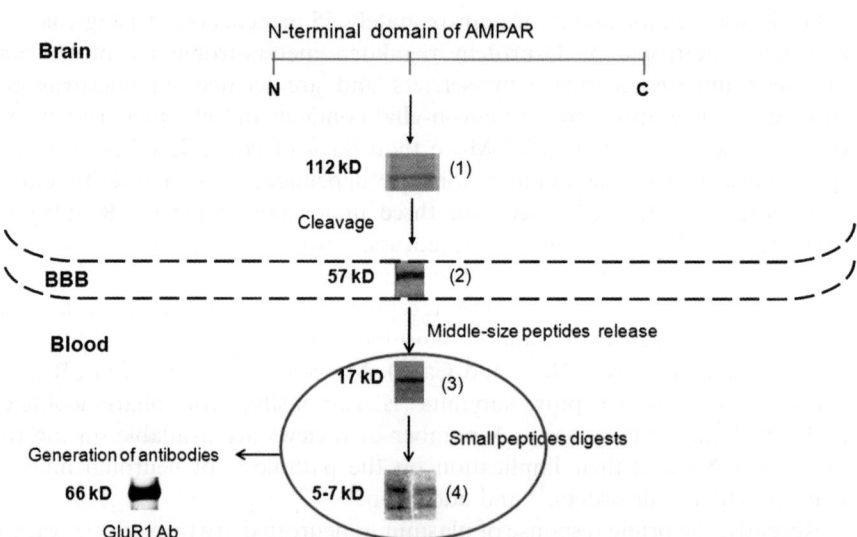

Figure 4.1   Proteolyses of AMPAR neuroreceptors. SDS-PAGE electrophoresis of quisqualate-binding protein isolated from synaptic membranes of post-mortem human cortex (1), Western blot of AMPAR proteolysis fragment in cortex (2), middle-size AMPAR peptides in plasma (3), AMPAR small peptides in plasma (4).

middle-size fragments (17 kD), and large-size fragments (57 kD). Fragments of 57 kD can be found in astroglial or microglial cells,[46] while 17 kD and smaller peptide fragments may penetrate the compromised BBB and enter the blood-stream, generating antibodies.[22]

CNS endogenous peptides that leak into the bloodstream generate an immune response.[10] Digested neuropeptides that contain only active immune epitopes (antigens) of 5–7 kD have been associated with induced cortical impact, head trauma, and spinal cord injury.[10,22] In addition, specific anti-bodies to neurotoxicity biomarkers are found to be present in significant quantities in peripheral fluids.[47]

Clinical studies have shown that ionotropic glutamate receptors provide real-time evidence of neurotoxicity and abnormal brain-spiking activity.[48] Mole-cular investigations of NMDA receptor and AMPA receptor peptides and antibodies found them to correlate primarily with cerebral ischemic events[49–51] and head trauma,[52–55] respectively. Degradation of kainite receptors detected in cerebrospinal fluid have been linked to spinal cord compression and incomplete injury.[32] Once proteolytic peptide fragments of kainite receptor are detectable in the bloodstream, increased peptide levels can be correlated with irreversible, secondary changes in spinal cord that cause movement disability.[56]

4.3.2 "Bottom Up" Approach

The practical application of using NR1, NR2, and AMPA receptor peptide fragments as endogenous antigens as prospective brain biomarkers was initially broached more than 20 years ago using both "bottom up" and "top down" strategies.[22] The "bottom up" process involves use of monoclonal antibodies raised to specific glutamate-binding proteins (phencyclidine [PhBP], quisqualate [QBP], and kainate [KBP]) to visualize the morphological location of NMDA, AMPA, and kainate receptors in normal and diseased nervous tissue,[46] with subsequent monitoring of specific peptide trafficking.[22,47] The "top down" approach is characterized utilizing molecular biological techniques using a cDNA library for identification of exact nucleotide sequences encoding peptides associated with certain disorders. Both approaches have been successfully uti-lized in the search for biomarkers of drug addiction[57–59] and epilepsy.[60,61]

Historically, both approaches have been initiated from isolation, purifica-tion, and study of properties of individual glutamate-binding proteins, including those in the glutamate receptor family. These studies involved various biochemical and immunochemical methods comprising affinity chromato-graphy, high-performance liquid chromatography purification, agonist and antagonist selective binding sites analyses, and functional properties of receptor fragments.[22] Most of the results from these studies have been published in non-English language professional journals; what follows is a summary of these data that relate to properties of ionotropic glutamate receptors.

Trafficking of proteolytic digestion products of quisqualate-binding protein in the blood of rodents and human has been demonstrated,[22,62] and purification of 17 kD and 57 kD AMPA receptor fragments from synaptic membranes of human cortex to which mono- and polyclonal antibodies have been raised has also been shown.[63–67] Maximum concentration of anti-AMPA mAb immunoreactivity was found in synaptic terminals of glutamate-receptive neuronal axons.[67,68]

Western blot analyses of human quisqualate binding protein (GBP) isolated from postmortem cortex using natural human IgG (A) and polyclonal hen antibodies (B) raised to synthetic AMPAR peptides recognized the same protein bands of 10–57 kD in the presence of protease inhibitors (see Figure 4.2). Incubation of GBP with proteinase K revealed protein band of 45 kD related to kainite binding proteins. Treatment with Pronase caused complete protein degradation of the amino acids (see Figure 4.2). Digested products obtained simultaneously by use of the nonspecific serine proteases trypsin and papain allowed detection of a single 17 kD protein band in both blots. These data confirm that original human antibodies and IgY raised to synthetic AMPA receptor peptide-labeled synaptic protein bands represent epitopes of native AMPA receptor protein.

In order to characterize QBP (AMPA receptor fragment), the protein was digested to obtain smaller molecular peptide fragments by proteolytic enzymes. Peptides were constrained (i) by trypsin specific to C-terminal arginine or lysine residues; and (ii) by thrombin-activated serine protease. Peptide fragments were analyzed by mass spectrometry, which allowed detection of a series of products originating from the quisqualate-binding fragments 17 kD and corresponding to approximately 60 peptides (see Figure 4.3A). Peptides were additionally

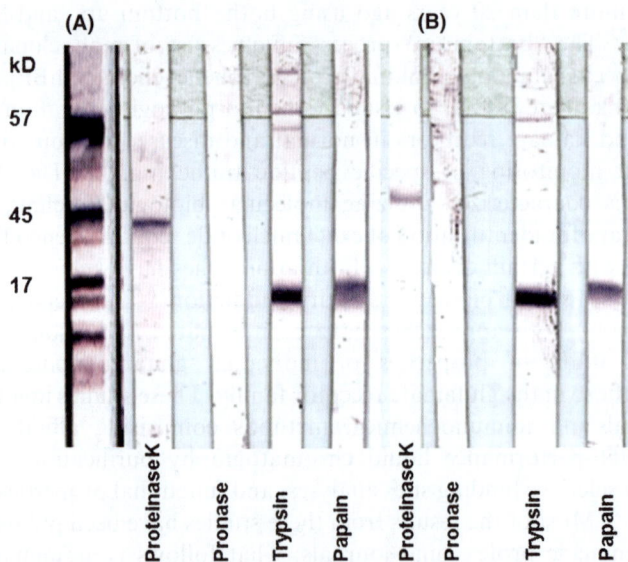

Figure 4.2   Western blot of human glutamate-binding proteins (2 μg) from postmortem cortical tissue with A – specific human IgG isolated from serum of TBI patient – and B – polyclonal IgY against AMPAR peptide. All digests were performed at 37 °C for 1 hour incubation.

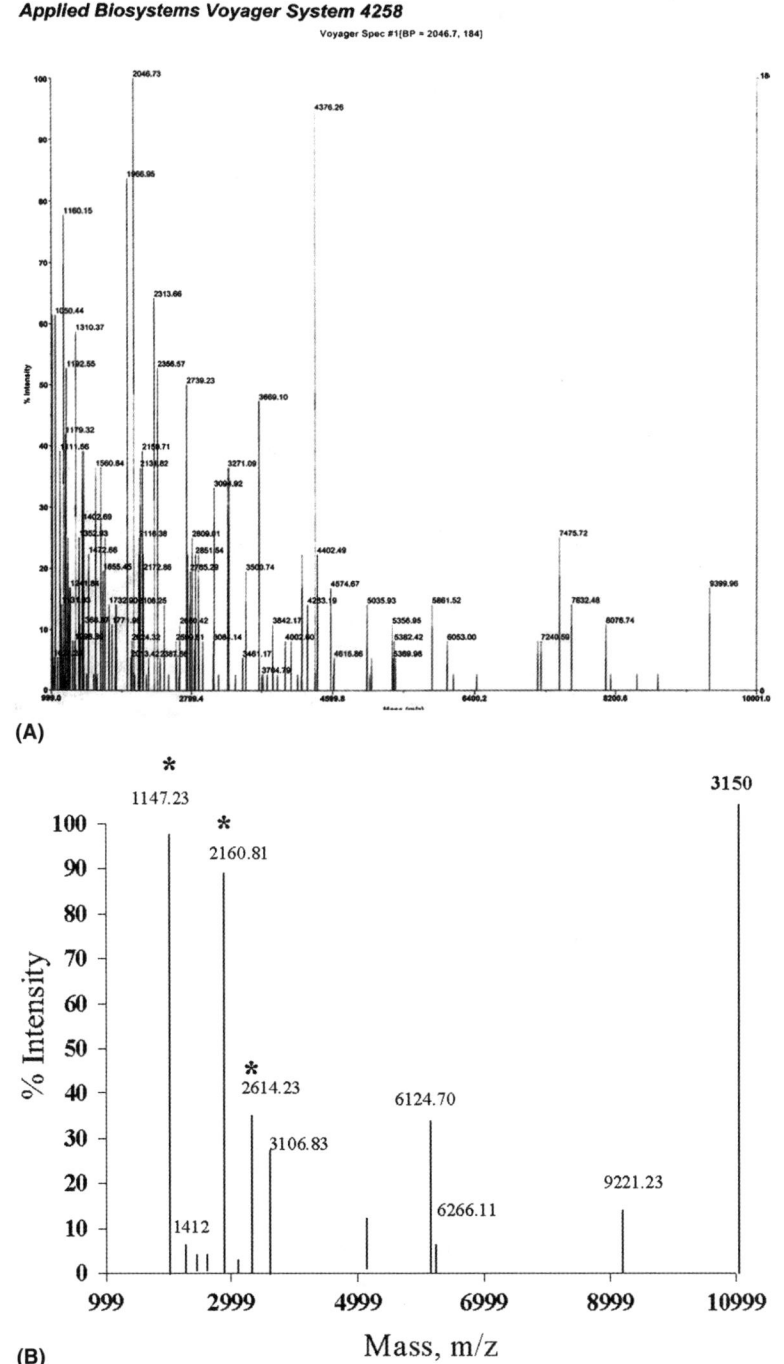

Figure 4.3   A: mass spectrometry of QBP trypsin digest. About 60 peptides with molecular weights 1–10 kD were obtained. B: reverse-phase HPLC purification using mAb. Stars show peptides reacted with mAb against AMPA receptors.

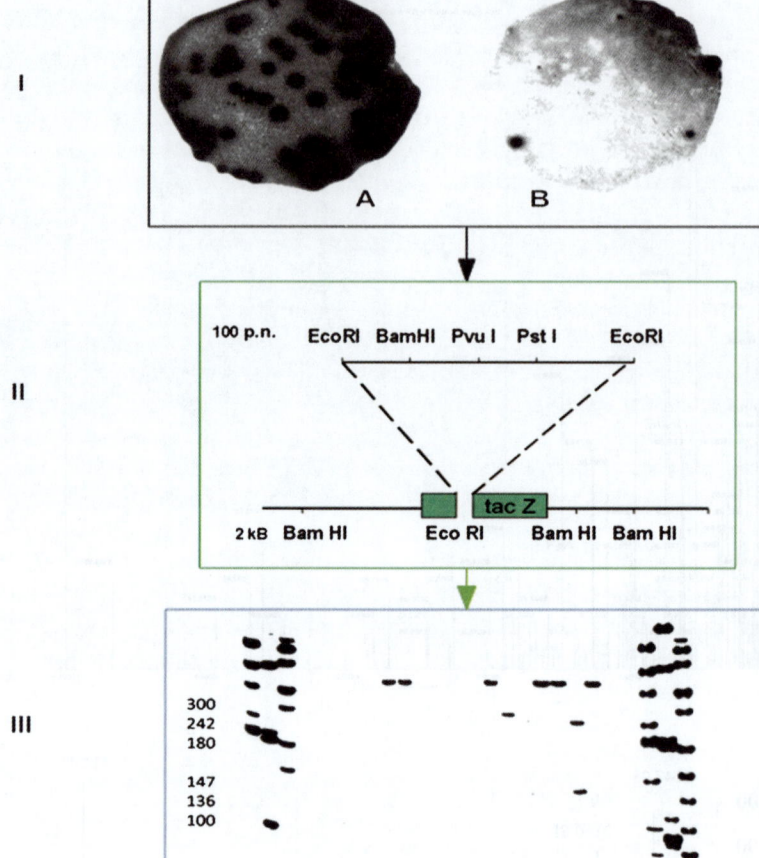

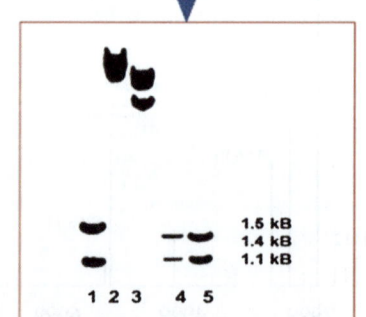

purified by reverse-phase HPLC and assayed with monoclonal (mAb) and polyclonal hen IgY antibodies against QBP. Mass spectrometry of labeled peptides demonstrated the existence of smaller peptides ($n = 3$) with molecular weights of 5–7 kD (see Figure 4.3B). All three isolated peptides were assayed against IgY isolated from serum of patients with post-traumatic epilepsy; the one possessing the most active epitope was selected as an endogenous antigen and prospective biomarker for brain injury.

4.3.3 "Top Down" Approach

Another approach to uncover novel biomarkers can be initiated by analyzing gene expressions of the targeted product. The characterization of little known membrane protein genes requires having minimal information concerning immunochemical properties and extra- or intracellular location. To define genetic fingerprints encoding these specific peptide fragments; for example, glutamate receptors or glutamate-binding proteins (GBP), a recombinant RNA or DNA library may be searched using poly- or monoclonal antibodies matched to the desired protein.[69]

In our research, we used purified GBP[63–67] to isolate IgG from pooled blood samples of patients with TBI or post-traumatic epilepsy. Immunoscreening cDNA library was then performed by raised mono-/polyclonal antibodies against GBP and isolated human IgG. The latter revealed several hybrid cDNA clones, which were incorporated as alien cDNA fragments in an *E. coli* bacterial system. This was then employed to express specific phages (λGT11) corresponding to desired products[70] (see Figure 4.4A).

Each of three λGT11-specific cDNA fragments was transferred into M13 plasmid vector and cloned. Large inserts of 450 kb to 690 kb were obtained corresponding to cDNAs for NMDA (λ-PhBP), AMPA (λ-QBP), and kainite receptors (λ-KBP). All three fragments were isolated from agarose gel after the cDNAs were treated with a number of standard nucleotide restrictase kits[69] (see Figure 4.4B). Three unique sites of the cDNA fragment (Pst1, BamH1, and Psal) of the larger products were revealed, and 5′-3′ oligonucleotide sequence orientation using Kpn1, BamH1, and EcoR1 was deduced for λ-PhBP, λ-QBP, and λ-KBP cDNAs.[71]

Figure 4.4 Immunoscreening cDNA library and λGT11 vector study. I: immunopositive signal from phage λGT11cDNA library labeled by IgG of patients with ischemic stroke. A: lGT11 control; B: lGT11 corresponding to phencyclidine-binding proteins (PhBP). II: EcoRI fragment of human PhBP cDNA restriction map. III: electrophoresis ^{32}P-λ PhBP cDNA fragment treated by restricting enzymes: 1: Hh1; 2: Msp1; 3: Tag1; 4: HindIII; 5: BlgII; 6: Bcl1; 7: BamH1; 8: Pst1; 9: Sal1; 10: EcoRV; 11: Rsa1; 12: untreated fragment; M1, M2, M3: length standards. IV: orientation of the λ PhBP cDNA: 1; 2: Kpn1 and BamH1; 3: Kpn1 and EcoR1; 4: DNA ladder; 5: length standards.

All isolated cDNA fragments coding human PhBP, QBP, and KBP were transcribed to mRNA and amplified by PCR. Northern blot analysis of ^{35}S-methionine-labeled products of all mRNA expressions was performed using polyclonal antibodies purified from serum of patients with ischemic stroke, TBI, or post-traumatic epilepsy. The main radioactivity bands corresponded to polypeptides of 68kD (PhBP), 57 kD (QBP), and 45 kD (KBP).[53]

After cDNA nucleotide sequence determination for each glutamate-binding protein, the open reading frame for three polypeptides, 17 kD (consisting of 164 amino acid residues [QBP]), 14 kD (containing 130 amino acids [PhBP]), and 10 kD (with 112 amino acids [KBP]), were found. The homologue cDNAs encoding the fragments of rat and human glutamate receptors were isolated and compared with known AMPA, NMDA, and kainate receptor sequences.[72,73]

A computer analysis of the known nucleotide sequence banks revealed the existence of 99%, 96%, and 89% of homologies for glutamate-binding proteins that corresponded to N-terminus domain of the NR2 subtype of NMDA receptors, the GluR1 subtype of AMPA receptors, and the GluR6/7 subtypes of kainite receptors.[73,74]

Discovery of sequences of peptide fragments resulting from proteolysis has allowed several synthetic peptides of different molecular weight from 2 to 5 kD to be synthesized. These peptides have been analyzed using earlier obtained mono- and polyclonal antibodies against native GBP (PhBP, QBP, and kainite-binding protein). Only peptides with high immunogenic activities to corresponding antibodies have been selected as potential neuroexcitotoxicity biomarkers for cerebral ischemia, head, and spinal cord injury.

4.4   Glutamate Receptor Biomarkers in Peripheral Fluids

Within the past two decades, the search for biomarkers to assess severity or consequences of brain injury has been directed to discovery of active antigens and antibodies. Hundreds of proteins have been isolated from the brain[75] and CSF after severe TBI.[76]

Primary injury to the brain or spinal cord causes mechanical damage and energy failure in parenchymal cells and endothelia that comprise the blood-brain and blood-spinal cord barriers. Traumatic CNS injury causes brain peptides to be released continuously into the bloodstream through the compromised BBB within hours to days after impact, while most proteins enter biological fluids relatively late, even weeks after the injury, due to breakdown of the BBB.[77] In general, initial BBB leakage is linked to trafficking of small or middle-size peptides that may have different lifetimes in blood and in CSF.[78] Some peptides can be rapidly digested by enzymes, whereas others with active epitopes can activate immune response and might serve as an evidence of asymptomatic or subclinical events.[79] The monitoring of these antigens in biological fluids allows one to define an early pathological cascade after acute brain injury.

Secondary injury mechanisms exacerbate tissue damage and BBB break-down. The combination of primary and secondary injury releases antigens (structural proteins) that can drain into secondary lymphoid tissues, thereby activating neuroantigen-reactive T and B lymphocytes and producing anti-bodies. There are more than 100 antibodies against known antigens, most of which belong to CNS proteins.[80]

One promising line of brain injury research would be utilization of both antigens and antibodies as biomarkers of neurotoxicity for assessment of TBI,[81,82] stroke,[10,22,34,47,79,83,84] and epilepsy.[85,86]

4.4.1 Natural Antigens and Antibodies Responses to Brain Damage

Nervous tissue damage, vascular injury, and BBB disturbance facilitate output of brain antigens released from injured neurons that can activate the immune system.[87] Although most antibodies and lymphocytes are held in check by complex mechanisms of central and peripheral tolerance, their normal immune potential can be unleashed when the BBB is compromised.[88] It is plausible that if produced after brain injury, antibodies may be considered "natural," and bind CNS antigens without causing harmful consequences, including infiltra-tion to the injured tissue.[89,90] Due to the nature of normal immune response, antibodies to natural antigens detected in serum are likely to be associated with advanced or chronic processes of brain damage,[79] while antigens measured in plasma could reflect acute injuries.[47]

A number of brain-borne antigens and antibodies that result from CNS injury have been identified. Antibodies that bind to N-terminus domain of NR2A/2B subunits of the NMDA receptor were found to increase significantly in people who suffered from transient ischemic attacks and acute ischemic strokes,[10,83] whereas antibodies to the AMPA receptor were registered in patients with post-traumatic epilepsy within a year after closed head injury.[22] Levels of NR2 antibodies in children with severe craniocerebral trauma (GCS < 9 points) were lower than in children with mild injury, with the lowest levels observed in the group who died following brain injury.[55]

Correct and timely detection of natural antigens for brain injuries may help in development of novel pharmaceutical or natural agents with neuroprotective properties. These should be administered immediately after impact to prevent subsequent progression of pathological injury. In addition, natural antibodies could assist in follow-up after treatment and aid in determining prognosis.[55]

4.4.2 Autoantibodies and Autoimmune Reactions to CNS Injury

Extensive reviews of brain injury and autoimmune reactions have suggested that anti-brain antibodies may play a role in phagocytosis and removal of damaged neurons.[91] When reactive lymphocytes and antibodies initiated by the combination of primary and secondary injury infiltrate injured tissue or are produced locally at the injury site, an autoimmune reaction is launched. Activation of neuroinflammatory (immunoexcitotoxic) cascades due to infection or other factors causes autoimmune diseases and progressive incurable disorders of immune system directed against autologous antigens.[91]

A hypothesis concerning the relationship between autoimmune response and immune system dysfunction generating autoantibodies to glutamate receptors for Rasmussen encephalopathy was previously reported.[92,93] Identification of antibodies to the GluR3 domain of AMPA and kainate receptors in the serum of patients with Rasmussen encephalitis, paraneoplastic neurological disorder, and other types of epilepsy supported this hypothesis.[85,92,94,95] However, these findings have been weakened by an inability to detect GluR3 antibodies in many patients with Rasmussen encephalitis who had pronounced symptoms.[96]

Recently, autoantibodies to GluR1/2 AMPA receptors have been described in patients with limbic encephalitis who responded well to immunotherapies.[97–99] Subsequently, autoantibodies to NR1[100] or splice variant NR1/NR2B[101] in patients with encephalitis, and NR2B autoantibodies in patients with Rasmussen encephalitis, chronic *epilepsia partialis continua*, and in acute limbic encephalitis were revealed.[102,103] Neuropathogenicity of GluR autoantibodies in systemic lupus erythematosus,[104] sporadic olivopontocerebellar atrophy,[105] and paraneoplastic encephalopathies[106] has also been found.

Some mental or emotional conditions may be related to specific autoantibodies circulating in abnormal amounts in blood of patients. Autoantibodies to dopamine receptors have been detected in serum of patients with schizophrenia, depression, and drug addiction.[107] Recently, researchers from Johns Hopkins School of Medicine (Baltimore, MD) demonstrated autoantibodies to gliadin and elevated tumor necrosis factor (TNF)-alpha in patients with recent-onset psychosis, multi-episode schizophrenia, and bipolar disorder, while autoantibodies to NMDAR appeared to be mania-specific.[108] Interventions using Cox-2 inhibitors or other anti-inflammatory agents and immunotherapy reduced levels of markers of immune activation.[109–111] In the future, similar sequences of known bacterial/viral pathogens and neuroreceptors should be explored to help explain cross-reactivity of antibodies; based on such an autoimmunity hypothesis, harmful infectious and non-infectious generated antibodies that may result in brain and spinal cord dysfunctions[91] can be isolated.

4.5 Conclusions

The search for specific biomarker(s) of a definite neurological disorder requires several alternative approaches, where "bottom up" and "top down" pathways might solve the challenge. The "bottom up" path involves proteolytic breakdown of CNS proteins and immune response to the digested products. The "top down" route focuses on genomic technology, where DNAs encoding peptides and proteins resulting from degradation are investigated using specific antibodies. The biomarker(s) discovered on the basis of both technologies might have potential for clinical use. Proteins are more likely to be studied by a complex of two-dimensional gel electrophoreses. This method may isolate proteins in very small amounts, which can be analyzed only using advanced mass-spectrometry methods.

Neurodegradomics as a common approach looks promising for searching for brain biomarkers associated with certain neurological deficits. Applied to chronic conditions, the detection of natural antibodies and autoantibodies will allow earlier diagnoses of nervous disorders and timely initiation of preventive treatment. The treatment strategy should utilize all of the complex steps of new biomarker(s) discovery, blood assay(s) development, test(s) feasibility study and clinical trials to validate the concept. Novel medication efficacy should be assessed by a panel of biomarker assays. Glutamate receptors as neurotoxicity biomarkers might play an important role in drug-test co-development.

References

1. C. M. Micheel and J. R. Ball, in *Evaluation of Biomarkers and Surrogate Endpoints in Chronic Disease*, The National Academy, Washington, DC, 2010, p. 22.
2. O. L. Alves and R. Bullock, in *Brain Injury (Molecular & Cellular Biology of Critical Care Medicine)*, ed. R. S. B. Clark and P. Kochanek, Kluwer Academic Publishers, London, 2001, p. 1.
3. S. F. Traynelis, L. P. Wollmuth, C. J. McBain, F. S. Menniti, K. M. Vance, K. K. Ogden, K. B. Hansen, H. Yuan, S. J. Myers and R. Dingledine, *Pharmacol. Rev.*, 2010, **62**, 405.
4. S. Gill and O. Pulido, in *Glutamate Receptors in Peripheral Tissue: Excitatory Transmission Outside the CNS*, ed. S. Gill and O. Pulido, Kluwer Academic Publishers, New York, 2010, p. 3.
5. B. Lomenick, R. W. Olsen and J. Huang, *ACS Chem. Biol.*, 2011, **6**(1), 34.
6. J. Laterra and G. W. Goldstein, in *Principles of Neural Science*, 4th edn, ed. E. R. Kandel, J. H. Schwartz and T. M. Jessell, McGraw-Hill, New York, 2000, p. 1288.
7. H. Reiber, *Clin. Chim. Acta*, 2001, **310**, 173.
8. H. Reiber, *Restor. Neurol. Neurosci.*, 2003, **21**, 79.
9. N. J. Abbott, *Cell. Mol. Neurobiol.*, 2000, **20**, 131.
10. S. A. Dambinova, G. A. Khounteev, G. A. Izykenova, I. G. Zavolokov, A. Y. Ilyukhina and A. A. Skoromets, *Clin. Chem.*, 2003, **49**, 1752.
11. D. W. Busija, in *Nervous Control of Blood Vessels*, University of Nottingham, CRC Press, London, 1996, p. 177.
12. C. Betzen, R. White, C. M. Zehendner, E. Pietrowski, B. Bender, H. J. Luhmann and R. W. Kuhlmann, *Free Radical Biol. Med.*, 2009, **47**, 1212.
13. C. D. Sharp, M. Fowler, T. H. Jackson, J. Houghton, A. Warren and A. Nanda, *BMC Neurosci.*, 2003, **4**, 28.
14. T. Zamalloa, C. P. Bailey and J. Pineda, *Br. J. Pharmacol.*, 2009, **156**(4), 649.

15. Y. Y. Jean, L. D. Lercher and C. F. Dreyfus, *Neuron Glia Biol.*, 2008, **4**(1), 35.
16. G. A. Davis, G. L. Iverson, K. M. Guskiewicz, A. Ptito and K. M. Johnston, *Br. J. Sports Med.*, 2009, **43**(Suppl. I), i36.
17. C. Werner and K. Englehard, *Br. J. Anesth.*, 2007, **99**, 4.
18. A. C. Rodriguez-Paez, J. P. Brunschwig and H. M. Bramlett, *Acta Neuropathol.*, 2005, **109**, 603.
19. P. K. Stys, *Curr. Mol. Med.*, 2004, **4**, 113.
20. S. B. Tekkok and M. P. Goldberg, *J. Neurosci.*, 2001, **21**, 4237.
21. S. B. Tekkok, Z. Ye and B. R. Ransom, *J. Cereb. Blood Flow Metab.*, 2007, **27**, 1540.
22. S. A. Dambinova, *Glutamate Neuroreceptors*, Nauka, Leningrad, 1988, p. 176 (in Russian).
23. S. A. Dambinova, in *Therapeutic Electrical Stimulation of Human Brain & Nerves*, ed. N. P. Bechtereva, Sova, St. Petersburg, 2008, p. 346.
24. R. L. Blaylock and J. Maroon, *Surg. Neurol. Inter.*, 2011, **2**, 107.
25. M. L. Block, L. Zecca and J. S. Hong, *Nat. Rev. Neurosci.*, 2007, **8**, 57.
26. B. Li, C. S. Piao, X. Y. Liu, W. P. Guo, Y. Q. Xue, W. M. Duan, M. E. Gonzalez-Toledo and L. R. Zhao, *Brain Res.*, 2010, **1327**, 91.
27. T. E. Golde, C. Zwizinski and A. Nyborg, in *Intramembrane-Cleaving Proteases (I-CLiPs)*, ed. N. M. Hooper and U. Lendeckel, *Proteases in Biology and Disease*, Springer, Dordrecht, 2007, p. 17.
28. H. Nakanishi, in *Proteases in the Brain*, ed. U. Lendeckel and N. M. Hooper, Springer, Dordrecht, 2005, p. 303.
29. S. M. Knoblach and A. I. Faden, in *Proteases in the Brain*, ed. U. Lendeckel and N. M. Hooper, Springer, Dordrecht, 2005, p. 79.
30. D. Allsop, J. Mayes, S. Moore, A. Masad and B. J. Tabner, *Biochem. Soc. Trans.*, 2008, **36**(Pt 6), 1293.
31. X. Bi, V. Chang, E. Molnar, R. A. J. McIlhinney and M. Baudry, *Lett. Neurosci.*, 1996, **73**(4), 903.
32. J. W. Olney, *Science*, 1969, **164**, 719.
33. D. W. Choi, *Neurosci. Lett.*, 1985, **58**(3), 293.
34. E. I. Gusev, V. I. Skvortsova, G. A. Izykenova, A. A. Alekseev and S. A. Dambinova, *Zh. Nevropatol Psikhiatr Im. S S Korsakova*, 1996, **5**, 68.
35. K. Pelkey and C. J. McBain, in *The Glutamate Receptors*, ed. R. W. Gereau IV and G. T. Swanson, Humana Press, Totowa, NJ, 2008, p. 179.
36. C. D. Sharp, M. Fowler, T. H. Jackson, J. Houghton, A. Warren and A. Nanda, *BMC Neurosci.*, 2003, **4**, 28.
37. D. E. Bergles, R. Jabs and C. Steinhäuser, *Brain Res. Rev.*, 2010, **63**(1–2), 130.
38. M. C. Ashby, M. I. Daw and J. T. R. Issac, in *The Glutamate Receptors*, ed. R. W. Gereau IV and G. T. Swanson, Humana Press, Totowa, NJ, 2008, p. 1.
39. R. S. Petralia and R. J. Wenthold, in *The Glutamate Receptors*, ed. R. W. Gereau IV and G. T. Swanson, Humana Press, Totowa, NJ, 2008, p. 45.
40. X. T. Jin and Y. Smith, in *Kainate Receptors*, ed. A. Rodriguez-Moreno and T. S. Sihra, Springer, New York, 2011, p. 27.

41. T. W. Lai, W. C. Shyu and Y. T. Wang, *Trends. Mol. Med.*, 2011, **17**(5), 266.
42. L. Garey, *J. Anat.*, 2010, **217**(4), 324.
43. M. E. Wolf, *Neurotox. Res.*, 2010, **18**(3–4), 393.
44. H. Yuan, K. M. Vance, C. E. Junge, M. T. Geballe, J. P. Snyder, J. R. Hepler, M. Yepes, C. M. Low and S. F. Traynelis, *J. Biol. Chem.*, 2009, **284**, 12862.
45. I. Schulte, H. Tammen, H. Selle and P. Schulz-Knappe, *Expert Rev. Mol. Diagn.*, 2005, **5**(2), 145.
46. S. A. Dambinova, Yu. V. Bobryshev, A. I. Gorodinsky, M. N. Margulis, A. O. Nikitin, E. A. Orlova, I. Y. Pavlov and A. Y. Savinov, *Zh. Evol. Biokhim. Fiziol.*, 1992, **28**, 201 (in Russian).
47. S. A. Dambinova, *Clin. Lab. Inter.*, 2008, **32**, 7.
48. S. A. Dambinova and G. A. Izykenova, *Zh. Vyssh. Nerv. Deiat. Im. I. P. Pavlova*, 1997, **47**(2), 439 (in Russian).
49. M. U. Gappoeva, G. A. Izykenova, O. K. Granstrem and S. A. Dambinova, *Biokhimia*, 2003, **68**, 696.
50. E. I. Gusev and V. I. Skvortsova, in *Brain Ischemia*, Kluwer Academic Publishers, Berlin, 2003, p. 151.
51. A. A. Skoromets, S. A. Dambinova, A. Yu. Iliukhina and V. A. Sorokoumov, *Zh. Nevrol. Psikhiat. Im. S.S. Korsakova*, 1997, **97**, 53.
52. M. M. Odinak, D. E. Dyskin, I. A. Toropov, A. Y. Emelin, A. A. Cherepanov and S. A. Dambinova, *Zh. Nevrol. Psikhiatr. Im. S.S. Korsakova*, 1996, **96**, 45.
53. S. A. Gromov, S. K. Khorshev, Yu. I. Poliakov, L. G. Gromova and S. A. Dambinova, *Zh. Nevropatol. Psikhiatr. Im. S.S. Korsakova*, 1997, **97**, 46.
54. A. V. Goryunova, N. A. Bazarnaya, E. G. Sorokina, N. Yu. Semenova, O. V. Globa, Zh. B. Semenova, V. G. Pinelis, L. M. Roshal and O. I. Maslova, *Neurosci. Behav. Physiol.*, 2007, **37**(8), 761.
55. E. G. Sorokina, Zh. B. Semenova, N. A. Bazarnaya, S. V. Meshcheryakov, V. P. Reutov, A. V. Goryunova, V. G. Pinelis, O. K. Granstrem and L. M. Roshal, *Behav. Physiol.*, 2009, **39**(4), 329.
56. N. P. Bechtereva, in *Magic of the Brain and Labyrinths of Life*, Sova, St. Petersburg, 2007, p. 56.
57. S. A. Dambinova and G. A. Izykenova, *Ann. N.Y. Acad. Sci.*, 2002, **965**, 497.
58. O. Granstrem, W. Adriani, D. Giannakopoulou, G. Izykenova, S. Dambinova and G. Laviola, *Neurosci. Lett.*, 2006, **403**, 1.
59. F. Capone, W. Adriani, M. Shumilina, G. Izykenova, O. Granstrem, S. Dambinova and G. Laviola, *Psychopharmacology (Berl.)*, 2008, **197**(4), 535.
60. T. Smirnova, J. Stinnakre and J. Mallet, *Science*, 1993, **262**, 430.
61. T. Smirnova, S. Laroche, M. L. Errington, A. A. Hicks, T. V. Bliss and J. Mallet, *Science*, 1993, **262**, 433.
62. S. A. Dambinova, A. I. Gorodinskii, T. M. Lekomtseva and O. N. Koreshonkov, *Biokhimia*, 1987, **52**, 1642.

63. S. A. Dambinova and V. I. Besedin, *Neurochemistry*, 1982, **1**, 352.
64. S. A. Dambinova and A. I. Gorodiinskii, *Biokhimia*, 1984, **49**, 67.
65. V. I. Besedin, A. S. Kuznetsov and S. A. Dambinova, *Biokhimia*, 1985, **50**, 363.
66. S. A. Dambinova and A. I. Gorodinskii, *Neurokhimia*, 1986, **5**, 115.
67. E. A. Orlova, G. F. Kalantarov, T. M. Smirnova and S. A. Dambinova, *Dokl. Akad. Nauk SSSR*, 1989, **307**, 495.
68. Y. V. Bobryshev, E. A. Orlova, I. V. Balabanova, A. S. Kuznetsov and T. M. Smirnova, *Cytology*, 1989, **31**, 176.
69. R. Rapley, in *Principles and Techniques of Biochemistry and Molecular Biology*, ed. K. Wilson and J. Walker, Cambridge University Press, Hong Kong, 2005, p. 225.
70. T. M. Smirnova, E. A. Orlova and S. A. Dambinova, *Dokl. Akad. Nauk SSSR*, 1988, **303**, 756.
71. S. A. Dambinova, *Antibody Assay Methods to Assess Risk for TIA/Stroke*, U.S. patent No. 7,658,911 B2, February 09, 2010.
72. T. M. Smirnova, E. A. Orlova and S. A. Dambinova, *Dokl. Acad. Nauk SSSR*, 1989, **309**, 745.
73. S. A. Dambinova, *Rapid Multiple Panel of Biomarkers in Laboratory Blood Tests for TIA/Stroke*, PCT patent No. WO 02012892, February 02, 2002.
74. S. A. Dambinova and G. A. Izykenova, *Immunosorbent Tests for Assessing Paroxysmal Cerebral Discharges*, PCT Application WO2005046442. Filed: 11/08/2004. Published: 05/26/2005.
75. S. Mondello, U. Muller, A. Jeromin, J. Streeter, R. L. Hayes and K. K. Wang, *Expert Rev. Mol. Diagn.*, 2011, **11**(1), 65.
76. J. Hanrieder, M. Wetterhall, P. Enblad, L. Hillered and J. Bergquist, *J. Neurosci. Methods*, 2009, **177**(2), 469.
77. S. Nag, A. Kapadia and D. J. Stewart, *Neuropathol. Appl. Neurobiol.*, 2011, **37**(1), 3.
78. L. Doeuvre, L. Plawinski, F. Toti and E. Anglés-Cano, *J. Neurochem.*, 2009, **110**(2), 457.
79. J. D. Weissman, G. A. Khunteev, R. Heath and S. A. Dambinova, *J. Neurol. Sci.*, 2011, **300**(1–2), 97.
80. D. Pleasure, *Arch. Neurol.*, 2008, **65**(5), 589.
81. S. A. Dambinova, *IVD Technology*, 2007, **3**, 15.
82. A. Shikuev, A. Bagumyan, U. Danilenko and S. Dambinova, *J. Neurotrauma*, 2011, **28**, A-55.
83. S. A. Dambinova, G. A. Khounteev and A. A. Skoromets, *Stroke*, 2002, **33**(5), 1181.
84. P. M. Bokesch, G. A. Izykenova, J. B. Justice, K. A. Easley and S. A. Dambinova, *Stroke*, 2006, **37**(6), 1432.
85. S. W. Rogers, P. I. Andrews, L. C. Gahring, T. Whisenand, K. Cauley, B. Crain, B. E. Hughes, S. F. Heinemann and J. O. McNamara, *Science*, 1994, **265**, 648.
86. S. A. Dambinova, G. A. Izykenova, S. V. Burov, E. V. Grigorenko and S. A. Gromov, *J. Neurol. Sci.*, 1997, **152**(1), 93.

87. Y. Zhang and P. Popovich, *Discov. Med.*, 2011, **11**(60), 395.
88. A. Basten and P. A. Silveira, *Curr. Opin. Immunol.*, 2010, **22**(5), 566.
89. M. R. Ehrenstein and C. A. Notley, *Nat. Rev. Immunol.*, 2010, **10**(11), 778.
90. B. R. Wright, A. E. Warrington, D. D. Edberg and M. Rodriguez, *Arch. Neurol.*, 2009, **66**(12), 1456.
91. S. Amor and R. Huizinga, *Autoimmunity to Neuronal Proteins in Neurological Disorders (Neurology-Laboratory and Clinical Research Developments)*, Novinka Books, Bel Air, CA, 2011.
92. R. E. Twyman, L. C. Gahring, J. Spiess and S. W. Rogers, *Neuron*, 1995, **14**(4), 755.
93. P. I. Andrews and J. O. McNamara, *Curr. Opin. Neurol.*, 1996, **6**, 673.
94. H. Wiendl, C. G. Bien, P. Bernaconi, B. Fleckenstein, B. C. E. Elger, J. Dichgans, R. Mantegazza and A. Melms, *Neurology*, 2001, **57**(8), 1511.
95. R. Mantegazza, P. Bernasconi, F. Baggi, R. Spreafico, F. Ragona, C. Antozzi, G. Bernardi and T. Granta, *J. Neuroimmunol.*, 2002, **131**(1–2), 179.
96. R. Watson, Y. Jiang, I. Bermudez, L. Houlihan, L. Clover, K. McKnight, J. H. Cross, I. K. Hart, A. Roubertie, J. Valmier, Y. Hart, J. Palace, D. Besson, A. Vincent and B. Lang, *Neurology*, 2004, **63**(1), 43.
97. M. Lai, E. G. Hughes, X. Peng, L. Zhou, A. J. Gleichman, H. Shu, S. Matà, D. Kremens, R. Vitaliani, M. D. Geschwind, L. Bataller, R. G. Kalb, R. Davis, F. Graus, D. R. Lynch, R. Balice-Gordon and J. Dalmau, *Ann. Neurol.*, 2009, **65**, 424.
98. L. Bataller, R. Galiano, M. García-Escrig, B. Martínez, T. Sevilla, R. Blasco, J. J. Vílchez and J. Dalmau, *Neurology*, 2010, **74**, 265.
99. F. Graus, A. Boronat, X. Xifro, M. Boix, V. Svigelj, A. García, A. Palomino, L. Sabater, J. Alberch and A. Saiz, *Neurology*, 2010, **74**, 857.
100. J. Dalmau, E. Lancaster, E. Martinez-Hernandez, M. R. Rosenfeld and R. Balice-Gordon, *Lancet Neurol.*, 2011, **10**(1), 63.
101. S. R. Irani, K. Bera, P. Waters, L. Zuliani, S. Maxwell, M. S. Zandi, M. A. Friese, I. Galea, D. M. Kullmann, D. Beeson, B. Lang, C. G. Bien and A. Vincent, *Brain*, 2010, **133**, 1655.
102. Y. Takahashi, H. Mori, M. Mishina, M. Watanabe, T. Fujiwara, J. Shimomura, H. Aiba, T. Miyajima, Y. Saito, A. Nezu, H. Nishida, K. Imai, N. Sakaguchi and N. Kondo, *Neurology*, 2003, **61**(7), 891.
103. Y. Mochizuki, T. Mizutani, E. Isozaki, T. Ohtake and Y. Takahashi, *Neurosci. Lett.*, 2006, **394**(1), 5.
104. C. Kowal, L. A. Degiorgio, J. Y. Lee, M. A. Edgar, P. T. Huerta, B. T. Volpe and B. Diamond, *Proc. Natl. Acad. Sci. USA*, 2006, **103**(52), 19854.
105. L. C. Gahring, S. W. Rogers and R. E. Twyman, *Neurology*, 1997, **48**(2), 494.
106. N. G. Carlson, L. C. Gahring and S. W. Rogers, *J. Neurosci. Res.*, 2001, **63**(6), 480.
107. P. Sokoloff, J. Diaz, B. Le Foll, O. Guillin, L. Leriche, E. Bezard and C. Gross, *CNS Neurol. Disord. Drug Targets*, 2006, **5**(1), 25.

108. R. Yolken and F. Dickerson, 6th, International Conference on Biochemical
     Markers for Brain Damage, 2011, **O**, 24.
109. F. Dickerson, C. Stallings, A. Origoni, J. Boronow and R. Yolken, *Prog
     Neuropsychopharmacol. Biol. Psychiatry*, 2007, **31**(4), 952.
110. F. Dickerson, C. Stallings, A. Origoni, C. Vaughan, S. Khushalani,
     A. Alaedini and R. Yolken, *Bipolar Disord.*, 2011, **13**(1), 52.
111. F. Dickerson, C. Stallings, A. Origoni, C. Vaughan, S. Khushalani,
     F. Leister, S. Yang, B. Krivogorsky, A. Alaedini and R. Yolken, *Biol.
     Psychiatry*, 2010, **68**(1), 100.

CHAPTER 5

Neurotoxicity Biomarkers in Experimental Acute and Chronic Brain Injury

ULIANA I. DANILENKO,[a] GERMAN A. KHUNTEEV,[a] ARTHUR BAGUMYAN[b] AND GALINA A. IZYKENOVA*[b]

[a] CIS Biotech, Inc., 2701 N. Decatur Road, Decatur, GA 30033, USA;
[b] GRACE Laboratories, LLC, Suite 212, 2675 N. Decatur Road, Decatur, GA 30033, USA
*Email: izykenova@yahoo.com

5.1 Introduction: Neurotoxicity Biomarkers in Experimental Brain Injuries

A number of studies have been devoted to experimental research of traumatic brain injury (TBI) affecting glutamatergic neurotransmission.[1-4] Impaired synaptic plasticity and excitotoxicity (neurotoxicity) with derangement in glutamatergic neurotransmission are among the processes triggered by TBI.[5-7] An acute increase in extracellular glutamate levels leads to overstimulation of glutamate receptors[7] through an increased influx of calcium into the cell, which is mainly mediated via specific pre- and post-synaptic glutamate receptor ion channels. Current research has demonstrated that ionotropic glutamate receptors play a key role in the mechanism of neurotoxic damage after experimental TBI.[1-10]

RSC Drug Discovery Series No. 24
Biomarkers for Traumatic Brain Injury
Edited by Svetlana A. Dambinova, Ronald L. Hayes and Kevin K. W. Wang
© Royal Society of Chemistry 2012
Published by the Royal Society of Chemistry, www.rsc.org

Similar to experimental stroke,[11] TBI produces immunologically active fragments of glutamate receptors (GluR antigens) that generate antibodies detectable in blood.[12] Accumulation of GluR1 antibodies in rat serum has been found to precede development of abnormal brain-spiking activity registered by EEG within 7 days after the brain injury. Rabbits immunized with a GluR3 peptide (245–457 amino acids) exhibited seizure-like behaviors, suggesting that antibodies to GluR3 may modulate neuronal excitability.[13] Attempts to use AMPA receptor (AMPAR) and NMDA receptor (NMDAR) peptides to reduce the effects of neurotoxicity in rat models of stroke and epilepsy have shown promising results;[14] with cortical lesion areas after brain injury significantly diminished. Glutamate receptor peptide fragments and antibodies may therefore be investigated for their utility in the diagnosis and treatment of brain injury.

In the search for biomarkers of TBI and treatment strategies for the neurotoxic consequences that follow brain injury, growing knowledge of molecular mechanisms underlying both the neuroprotective and neurodestructive effects of NMDAR agonists and antagonists has been accumulated.[15] Several major clinical trials in severely head injured patients that used NMDAR blockers failed to show efficacy or had to be stopped prematurely due to increase in mortality or worsening of the outcome.[16] To overcome these obstacles, it would be plausible to examine the natural peptide inhibitors of glutamatergic neurotoxicity underlying brain injury. These inhibitors might prevent neuronal death in response to both hyper- and hypo-activity of glutamate receptors underlying TBI.[17]

In this chapter, the utility of peptide fragments of ionotropic glutamate receptors detected in the blood to serve as antigens of brain injury is explored using experimental models of TBI. Also outlined are the potential effects of endogenous peptide inhibition of ionotropic glutamate receptors as preventive therapy for brain injury.

5.2 NMDA and AMPA Receptor Antigens after Experimental Concussion

NMDAR and AMPAR are key components that control the neurotoxicity cascade by engaging in immediate and secondary cumulative sequelae following TBI, with acute and chronic changes (see Chapter 4). On the basis of molecular, biochemical, and immunohistochemical investigations, the AMPAR fragment peptide was proposed as a biomarker of initial necrotic events and the NR2 peptide, an NMDAR fragment, as a biomarker of secondary ischemic events following acute mild brain injury.[18,19] Alterations of AMPA and NR2 receptor subunit expressions, densities in the brain, and subsequent trafficking of degradation products in the bloodstream through the compromised blood-brain barrier (BBB) after mild cerebral concussion injury were investigated.

5.2.1 Effect of Concussion Injury on Behavior and Neurological State

Animal models of TBI include focal, diffuse brain injury, coma, and their combinations.[20] In the most often used focal model of weight drop or controlled cortical impact, variable weights (20–75 g) and length of free fall on an exposed rodent skull produces brain injury of differing severity. However, controlled cortical impact models cannot fully reproduce the entire heterogeneous and multifaceted spectrum of clinical mild TBI. In addition, since these models invariably require a craniotomy, some pathophysiological aspects of mild closed head injury might not be accessible.

To resemble human concussion injury more closely, mouse models developed to study the pathophysiology of closed head mild TBI[21,22] have been adjusted for rats.[23] Detailed physiological, behavioral, and brain imaging data for this rodent model have been described; however, there are to date no investigations on the effects of endogenous peptide inhibition of glutamate receptors on neurotoxicity.

Mild cerebral concussion injury (mild CCI) by the weight-drop method[23] was induced in Sprague-Dawley rats ($n = 69$)[*] lightly anesthetized with halothane. A cylindrical brass weight (170 g) was freely dropped to strike a cylindrical polyacetal transducer rod that was positioned in the Plexiglas tube (11 mm in inner diameter, 1 m length) to 30% of its length and placed with its spherical tip directly on the rat's skull posterior to the bregma under the tip in vertical position. The animal's head was immobilized laterally by a sponge allowing some anterior–posterior motion, but no rotational head movement at the moment of impact. Sham rats ($n = 30$) were anesthetized, surgically prepared, and placed under the impact device, but were not subjected to injury.

Tests conducted on righting times to assess the latency period for each rat to right itself from the supine position after the termination of anesthesia (pairwise comparisons) showed that there was a significant main effect of impacted group ($p < 0.001$, ANOVA). The impacted rats took significantly longer to right themselves (5.13 ± 1.8 min) compared with the sham controls (3.34 ± 0.5 min), displaying a prolonged recovery period.[23]

At 24 hours, the rats were evaluated on the Morris water maze.[24] This task involved training each animal to navigate to a platform in a pool (200 cm wide × 40 cm deep) of opaque water to escape out of the water. After the 8th place condition block of trials, performance levels in impacted and sham groups began to gradually improve over time, although the sham controls generally performed at higher levels until the last block of trials (Figure 5.1A). The impacted rats had significantly longer mean escape path lengths ($p = 0.021$) and mean latencies ($p = 0.030$) compared with sham controls, in concurrence with previous findings.[23]

---

[*]The protocols for animal use and care have been approved by the Institutional Animal Care Committee at Pavlov's State Medical University and Institutional Animal Care and Use Committees at Emory University (Atlanta, GA).

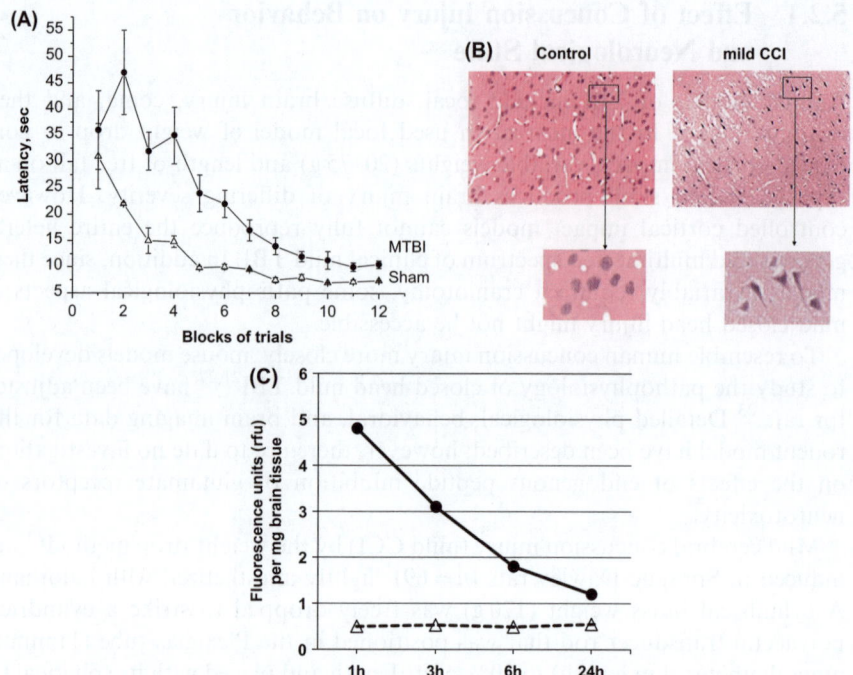

Figure 5.1   A: results of traditional Morris water task training of latencies to reach the platform grouped by blocks of two trials in rats with mild controlled cortical impact (black dots, $n=6$) and sham animals (white triangles, $n=6$). B: histochemical Nissl staining of 2 mm coronal slices from mild controlled cortical impact (CCI) rats ($n=6$) and sham animals ($n=6$; control). Impacted cortical area is shown. C: effect of mild CCI on BBB permeability. Fluorescence was measured in each hemisphere after mild CCI and 30 minutes after Na-fluorescein injection in rats with mild CCI (black dots, $n=3$) and sham animals (open triangles, $n=3$).

The severity of neurological symptoms in rats with mild CCI was graded at 24 hours post cortical impact, as described previously.[25] Experimental animals did not show circling (grade 3, severe) symptoms, decreased resistance of the forearm to lateral push (grade 2), or forelimb flexion (grade 1). In this study, rats with cortical impact were behaviorally normal and characterized by grade 0 from the Bederson's scale as found in other studies with large numbers of neurological tests performed.[22,23,26]

5.2.2   Assessment of Brain Damage and Blood-Brain Barrier (BBB) Integrity

To assess effects of experimental mild concussion injury, morphological changes in impacted cortical areas and BBB integrity were investigated. Frontal brain slices of mild CCI rats (5 μm) were stained by Nissl to assess the neuronal alterations in the impacted area. Induced mild CCI resulted in local neuronal damage presented as visible changes of neuronal shapes from round in the contra-lateral area of cortex (control) to stretched-out pyramidal structures in the injured area (see Figure 5.1B).

The measurement of fluorescence intensity 30 minutes after Na-fluorescein injections at 1, 3, 6, and 24 hours after cortical impact demonstrated that brain injury resulted in BBB leakage (see Figure 5.1C). BBB permeability increased by 1.7-fold at hour 3 and reached maximum at 24 hours post injury. In the post-mortem hemispheres of MCAo rats, the levels of fluorescence units were 5 times higher than those in sham-operated animals, indicating a compromised BBB.

Neuronal shrinkage was registered in impacted cortex at 24 hours after trauma, similar to that reported in an earlier study, in which changes in neuronal shapes in the central portion of the supraventricular cerebral cortex in mice were observed. No remarkable tissue disorganization was reported in Nissl-stained rat coronal sections after mild CCI depicted on the ninth day after injury.[23] However, a significant reduction of neurons in hippocampal sectors CA1 and CA3 was detected.[23]

The use of highly sensitive fluorescent marker allowed registration of BBB leakage not previously detected in studies of rodent models of mild TBI.[22,23] This discrepancy might be explained by the somewhat lower sensitivity of the Evans Blue stain used in assessment of BBB integrity in mice with mild CCT[23] and traditional MRI scanners (1.5–2.35 Tesla) utilized in both studies in order to detect structural integrity.[23] Traditional 3.0 Tesla MRI scanners take images of the brain based predominantly on signals emanating from water molecules and when actual edema has not yet developed, low resolution signals cannot differentiate structural changes.

5.2.3 Cortical NMDA and AMPA Receptor mRNA Expression and Receptor Subunit Densities

Protein synthesis in the injured brain changes significantly and is usually temporary; therefore, alterations in protein expression that underlie mild TBI may be reversible within a limited period of time.[27,28] To determine whether biomarkers of mild TBI may exist, the time course of AMPAR and NMDAR syntheses and protein density alterations in damaged cortical area after mild CCI were explored.

Expression of NR2 mRNA in the impacted cortex increased, with a maximum at 24 hours (92%), then significantly decreased by 72 hours (59%) after injury (see Figure 5.2A). Initial up-regulation of GluR1 mRNA expression immediately after cortical impact was followed by suppression (see Figure 5.2A); at 24 and 72 hours after mild CCI. GluR1 mRNA was significantly reduced compared with the control (see Figure 5.2B).

NR2 and GluR1 mRNA expression was not detectable in peripheral mononuclear cells at any time point after mild CCI or sham control rats (see Figure 5.2C). At the same time points, the standard GAPDH mRNA expression was readily detected in mononuclear cells.

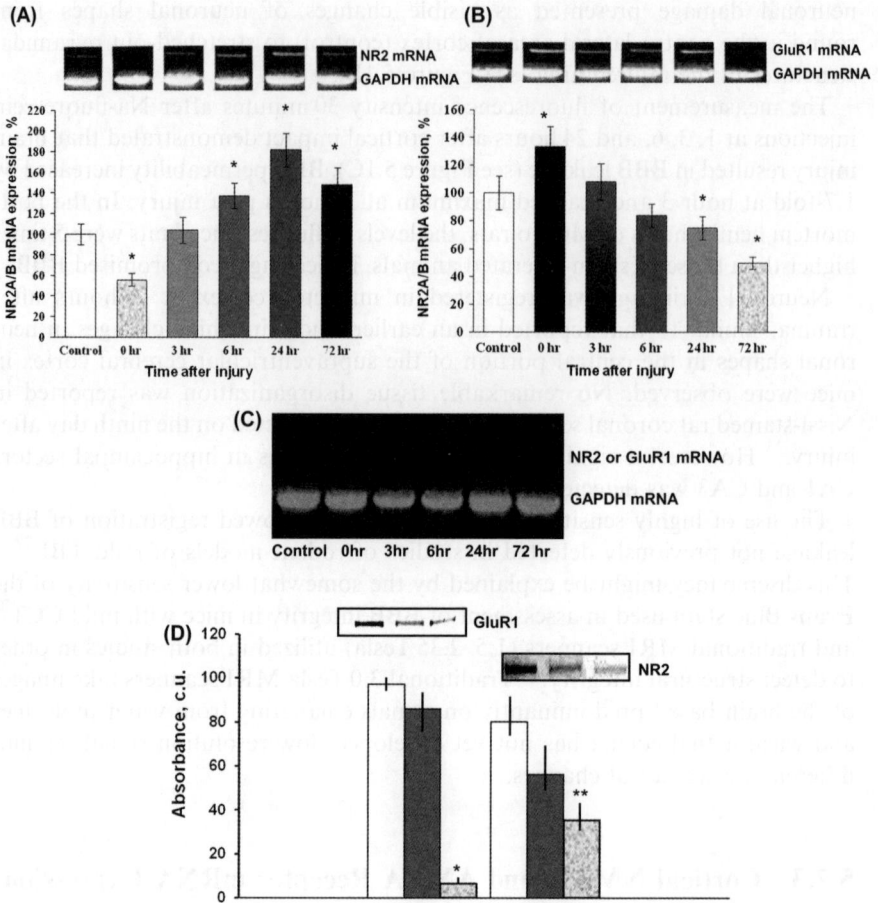

Figure 5.2 RT-PCR results in rat cortices after mild controlled cortical impact
(CCI). A: NR2 mRNA expression initially down-regulated immediately
after the injury (0 hour) followed by an up-regulation of 12% at 2 hours
and of 35% at 6 hours after mild CCI. At 24 and 72 hours after impact,
NR2 mRNA levels remained significantly increased at 92% and 59%,
respectively ($p < 0.001$, $F(5,30) = 36.9$, one-way ANOVA; $p < 0.05$,
Dunnett's test). B: GluR1 mRNA expression was significantly increased
immediately after acute injury (33% above the control at 0 hour) and
then gradually declined by 72 hours after mild CCI (*$p < 0.05$, one-way
ANOVA). C: no detectable NR2 and GluR1 mRNAs expressions
were found in the white blood cells of rats with mild CCI. All PCR
amplifications were assessed relatively to a glyceraldehyde-3-phosphate
dehydrogenase (GAPDH) standard (600 bp). D: GluR1 and NR2
peptide immunoreactivities in cortexes from 1, sham animals, 2, rats
with mild CCI at 0 hour after injury, and 3, rats with mild CCI at
24 hours after injury. Data presented as mean ± SD (*$p < 0.05$, one-way
ANOVA).

Thus, both NR2 and GluR1 mRNA expressions were decreased in impacted cortical site on the third day after mild CCI. Previously, it was demonstrated that frontal cortical ablation caused the decline in expression of normally abundant NMDAR (NR1, NR2A, NR2B) and AMPAR (GluR1-4) in deafferented dorsolateral striatum beginning at day 3 after injury.[29] Additionally, diffuse brain injury (DBI) was found to alter differently expression of groups I–III metabotropic glutamate receptors (mGluRs).[30] In comparison with that in the normal and sham control, expression of mGluR4, 7, and 8 (group III) increased significantly 1 hour after injury and peaked at 6 hours after DPI, while the number of group I mGluR positive neurons detected were elevated, with a delay at 12 hours and peaked at 24 hours. The concurrent expression pattern of group II mGluRs had exactly the opposite tendency to group I mGluRs, yielding significant decline by 24 hours of DBI.

Western blots of protein samples isolated from the impacted cortical site at 0 and 24 hours after brain injury as well as from sham control animals using monoclonal IgG against GluR1 and NR2 peptides were performed. In impacted cortex, GluR1 and NR2 peptides immunoreactivities were reduced by 11% and 24% at 0 hours after the injury, respectively (see Figure 5.2D). There was further reduction of densities for GluR1 peptide by 91% ($p < 0.001$) and for NR2 peptide by 45% ($p < 0.001$) in the impacted cortical site at 24 hours after the injury. These results are consistent with those of a prior study demonstrating that experimentally induced closed head injury decreased densities of activated NMDARs and AMPAR GluR1 subunits in the cortical impact site.[4] In the hippocampus, both receptor fragment densities were found elevated.

Thus, despite increased NMDAR and AMPAR syntheses in the impacted cortical area, GluR1 and NR2 protein densities are significantly depleted by 24 hours following mild CCI. This suggests that GluR1 and NR2 peptide fragments escape the brain through the compromised BBB and enter the bloodstream, where they may be detected and serve as potential biomarkers.

5.2.4 Proteolytic Receptor Fragments in Peripheral Fluids

The current neuroproteomic approach in the search for biomarkers of acute TBI employs mapping protein changes after an injury.[31–33] Protein fragments detected in cerebrospinal fluid (CSF) or blood are usually compared against existing peptide libraries to reveal prospective antigen(s).[33] Protein fragments with the highest immunogenicity are regarded as possible antigens and require validation as potential biomarkers utilizing TBI models.

Similar to stroke,[34,35] acute TBI and spinal cord injury (SCI) produces antigens that may be detected in CSF[32,33] and blood.[34] More than 60 cytoskeletal and other protein proteolytic fragments have been discovered that are released from rat cortical neurons after injury;[36] approximately 42 fragments have potential utility as protein biomarkers for TBI.[37]

However, the proteomic approach has several limitations and, to date, ongoing research has yet to show sensitive and reliable biomarker(s) for TBI. In the quest for mild TBI biomarkers, a few main issues need to be addressed – the first is time since injury.[18] Blood samples should be drawn as earlier after injury as possible so that essential biomarkers are not lost to digestive enzymes. In addition, while both CSF and blood undoubtedly contain traces related to the CNS, not all proteins found in CSF cross into the bloodstream; the CSF contains up to 76% proteins that are unique to this fluid.[38] Therefore, use of the correct experimental model of mild TBI would help address some of the current challenges. To understand the impact of brain injury on primary and secondary injury sequelae, polyclonal and monoclonal antibodies were used to monitor corresponding GluR1 and NR2 peptides in plasma of animals within 90 days after mild CCI. Significant amounts of GluR1 peptide (2 kD) were shown to enter the bloodstream within first 3 hours after the impact. In rats with mild CCI, only GluR1peptide blood level increased three- to fourfold within 6 hours after the impact compared to that for controls (see Figure 5.3).[39] Elevated amounts of GluR1 peptide were maintained in rats with mild CCI up to 3 months.

A low molecular weight NR2 peptide (2.5 kD) was detected in the same samples and indicated insignificant increase at 3 and 8 hours post injury compared to sham controls at baseline. The NR2 peptide reached maximum level at 24 hours and gradually declined by 14 days after mild CCI and remained insignificantly increased within 3 months of mild injury ($p < 0.05$, two-way ANOVA).

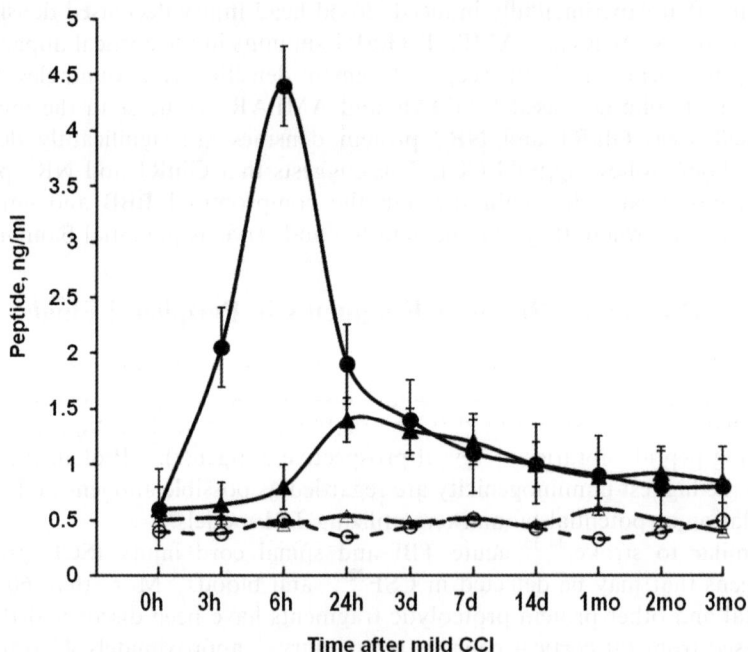

Figure 5.3   AMPAR (black circles) and NR2 (black triangles) peptide profiles in rats after mild controlled cortical impact (CCI) and respective sham controls, open circles for AMPAR, open triangles for NR2 peptide; $n = 6$ rats per group. Data presented as mean $\pm$ SD ($p < 0.001$; two-way ANOVA).

Abnormal amounts of NR peptide detected in the blood have been associated with cerebral ischemic complications;[18,34] therefore, slight changes in NR2 peptide measured in rats with mild CCI may be explained by development of secondary ischemic events surrounding area of impact, with a delay of 24 hours. This assumption should be investigated further with high-resolution MRI scans. GluR1 peptide was present in the blood of experimental animals in abundant amounts within 3 months after mild CCI. Therefore, the question arose, could GluR1 peptide have immunoactive epitope(s) and be a potential biomarker of mild TBI? The response might be positive if this peptide can produce an immune response.

5.3 AMPA Receptor Antibodies in Chronic Consequences after Experimental Concussion

Several subsequent events may occur in nervous tissue: (1) regions of undamaged hippocampus enriched of GluR1-4 subunits of AMPAR may be expressed differently in order to compensate for cortical injury; (2) peptide fragments of AMPAR containing immunologically active epitopes of N-terminus domain may be released through the compromised BBB into the bloodstream; and (3) the immune system recognizes the peptide as a foreign antigen and responds by generating antibodies.[12,38]

5.3.1 Distant Effect of Concussion Injury

No structural changes were observed in brain slices from rats with mild CCI at 24 hours after mild injury. However, MRI images performed in rats with mild CCI by day 9[23] showed neuronal reduction in hippocampal sectors CA1 and CA3 saturated with GluR1 and GluR2/3 subunits of AMPAR. To assess brain damage development, TTC-stained coronal sections obtained on day 14 following mild CCI were tested and found to have varying degrees of damage. There was some slight color fading in the swollen area of the cerebral cortex ($n = 6$) at the weight-drop site without observable damage. The volume of abnormality, mainly limited to the upper layers of the cortex in all slices (each 2 mm thick), was $15.1 \pm 2.2 \, \text{mm}^2$. No visible damage was found in the hippocampus in rats with mild CCI, while in a prior study, neuronal loss in hippocampal sectors was detected by the ninth day after injury.[23] There was visible partial cell loss in the impacted area without necrotic cavity but with area of damage similar to edema. These distant structural changes 14 days after mild CCI were similar to that obtained earlier using cortical impact and TTC staining of brain slices at 24 hours after impact.[26]

5.3.2 AMPA Receptor Subunit mRNA Expression and Protein Densities in Hippocampus

Assessment of GluR1 and GluR2/3 receptor mRNA expressions on day 14 after mild CCI showed up-regulated expression of both GluR1 and GluR2/3

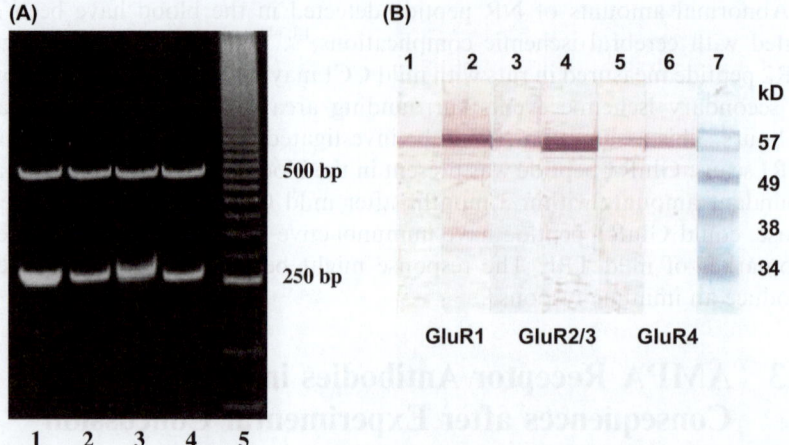

Figure 5.4   A: RT-PCR of GluRl and GluR2/3 RNAs (250 bp) in hippocampus from
rats on 14th day after mild controlled cortical impact (CCI). GluR1
mRNA expression: 1: animals with mild CCI; 2: sham controls. GluR2/3
mRNA expression: 3: animals with mild CCI; 4: sham controls. The
molecular ladder (5) and the expression of control actin mRNA (500 bp)
are shown. B: Western blot of rat hippocampus samples with polyclonal
antibodies against GluR1-4: (1, 3, 5): from sham controls, $n = 6$; (2, 4, 6):
from rats with mild CCI, $n = 6$; samples were collected on day 14 after
surgery; (7): molecular markers.

mRNA in the hippocampus adjacent to the impacted cortical site with constant
expression of the control β-actin mRNA (see Figure 5.4A). The growth in
AMPAR expression in the hippocampus most likely was directed to compen-
sation of the developing injury in the impacted cortical site.

The same hippocampal samples were used for assessment of immunoreactivities of GluR1, GluR2/3, and GluR4 subunits of AMPA receptor using
corresponding polyclonal antibodies (see Figure 5.4B). The significant increase
of GluR1 and GluR2/3 were detected in rats with induced mild CCI compared
to that for sham animals. There were no changes registered for GluR4
immunoreactivity.

Immunohistochemical staining of hippocampal slices with monoclonal
antibodies against GluR1 and *in situ* hybridization with nucleotides encoding
N-terminal fragment GluR1 (2–2.1 kD) have been performed on day 14 and 3
months after mild CCI. The GluR1 immunoreactivity increased mostly in the
CA3 region of rats with mild TBI on day 14 (see Figure 5.5). By the third month
after surgery, both immunostaining and *in situ* hybridization showed the
decline in amount of CA3 neurons. Dissipation of GluR1 deposits by the third
month coincided with abnormal brain spiking activity registered by EEG.
Convulsions in 30% of animals with mild CCI have been observed by 90 days.

Several questions remains to be addressed: is increased expression of
AMPAR receptor subunits in the hippocampus associated with developing
injury in the cortex? Does mild CCI produce GluR1 peptide with active

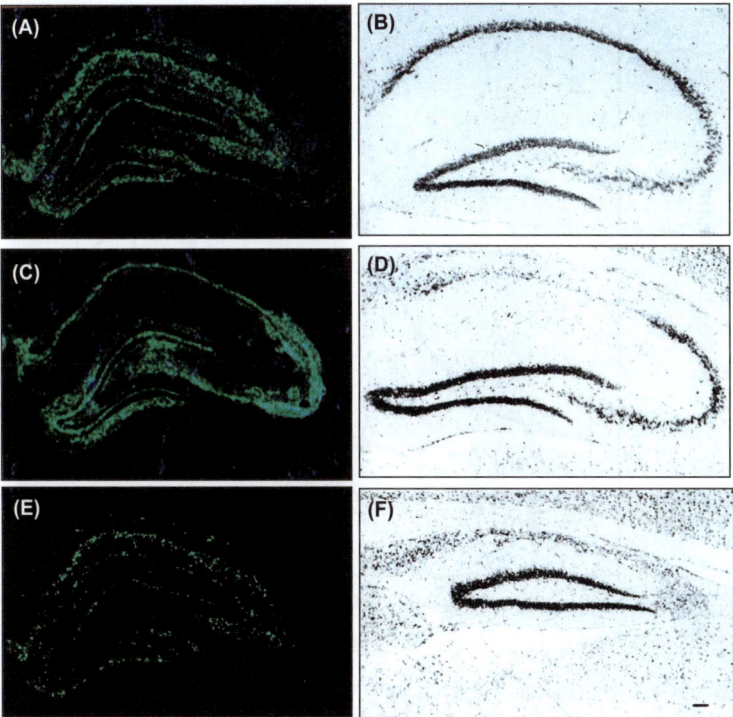

Figure 5.5 Immunostaining of hippocampal slices with GluR1 antibodies (A, C, E) and *in situ* hybridization (B, D, F). A, B: sham-operated rat; C, D: on day 14 of mild controlled cortical impact (CCI); E, F: 3 months after mild CCI. All *in situ* hybridizations were performed by use of nucleotide fragment encoding GluR1 (2–2.1 kD).

immune epitopes that, after entering the blood, generates antibodies? If glutamate receptor antibodies in detectable amounts circulate in the bloodstream, would their accumulation be predictive of secondary complications, such as abnormal brain spiking activity and convulsions or ischemic complications?

5.3.3 Detection of AMPA Receptor Subunit Antibodies in Serum

To explore the hypothesis that digestion products of GluR1 protein may be a biomarker of mild TBI, a synthetic GluRl peptide, the amino acid structure of which was derived utilizing a "top-down" neurodegradomic approach (see Chapter 4), was used to monitor GluR1 antibodies (Ab) in blood sera of rats with mild CCI and control animals.

Levels of GluR1 Ab in serum of rats with mild CCI were found to increase significantly beginning 24 hours after the injury compared to that for control animals (see Figure 5.6). These amounts reached the highest accumulation in

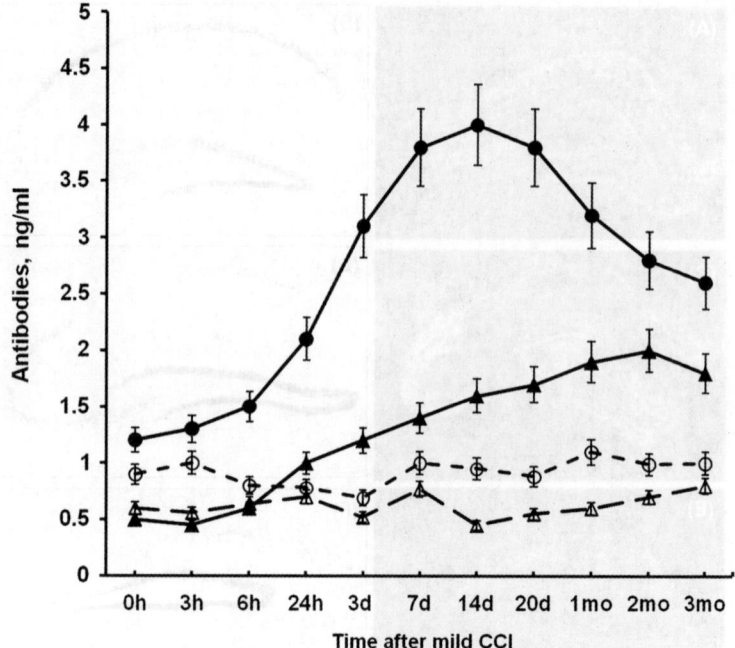

Figure 5.6   AMPAR (black circles) and NR2 (black triangles) antibody (Ab) profiles
in rats after mild controlled cortical impact and respective sham controls;
open circles for AMPAR, open triangles for NR2 antibodies, $n = 6$ rats
per each group. Data presented as mean $\pm$ SD ($p < 0.01$, two-way
ANOVA).

the serum of experimental rats to day 14 and remained significantly elevated
up to 3 months of observations, coinciding with increased GluR1 mRNA
expression, GluR1 immunoreactivities in hippocampus, and development of
abnormal brain spiking activity. The GluR1 Ab in serum of control rats
remained approximately 1.0 ± 0.02 ng/ml within the 3-month monitoring period. Simultaneously, NR2 Ab steadily grew and peaked around the second
month (day 60) after the mild injury and maintained elevated above the control
levels by day 90 of study in 50% of animals with experimental concussion. This
may be associated with edema-like formation in the impacted cortical area and
connected to secondary ischemic complications.

Immune reactions directed here can be viewed as a response to natural brain
antigen or self-directed adaptive immune response reactions.[40] The latter
involves robust activation of myeloid cells and then activation of self-reactive
lymphocytes with enhanced synthesis of antibodies.[41] Each component has
been implicated in progression of secondary injury after CNS injury.[42–46]

It is possible that if most antibodies produced after CNS injury are "natural"
antibodies, they could bind CNS antigens without causing harm.[47,48]
Mechanistic studies indicate that by binding glycolipids and proteins on glia

and neurons, natural antibodies trigger intracellular signaling pathways that promote protection and repair.[48,49] Experimental data provided evidence that NMDAR1 antibodies can be protective in stroke and epilepsy.[14] However, the study of mild CCI, cobalt-induced injury,[50] and independent research[51] all have shown that accumulation of natural antibodies can exacerbate brain-related convulsions[50,52] and ischemia/reperfusion injury.[51]

Current clinical study results based on antibodies for other neurological conditions are indicating their important role in disease progression and worsening condition.[13,34,38,52,53] However, further research will be needed before the significance of natural antibodies is understood.

5.4 Effects of Endogenous Peptide Inhibitor of AMPAR on Cortical Impact

Despite results of studies of experimental models of head injury that demonstrated glutamate antagonists to be neuroprotective,[3,10,15,16] several major clinical trials in patients with TBI failed to show efficacy.[16] A new prospective "avenue" in the search for more advantageous neuroprotective therapy may therefore be to explore endogenous peptides which, if used for TBI treatment, will not yield in harsh adverse events. Analogous to stroke,[54,55] neuroprotective mechanisms may induce tolerance through endogenous peptide agents when the brain is exposed to TBI. A natural peptide regulator, cortexin, has shown efficacy in treatment of patients with stroke[56] and severe TBI;[19] however, a better approach to treat TBI sequelae may be to look for a natural peptide inhibitor of AMPAR with moderate blocking power that would protect the brain from deterioration.

5.4.1 Endogenous Peptide Factor of Glutamate Receptors

The neuroprotective strategies for TBI sequelae should be related to prevention/inhibition of the glutamatergic system. It was hypothesized that the presence of specific endogenous peptide factor (EPF) inhibitor might regulate signal transmission in glutamatergic synapses.[57]

EPF fraction from rat synaptic membranes was isolated and revealed that the factor consisted of natural peptides with molecular weights of 450–600 Da.[17] The purified peptide sequence contained double glutamate derivates. The EPF competed for ^{3}H-L-glutamate binding to synaptic membranes with inhibition constant (K_i) of 4.8, demonstrating reversible blockage compared to irreversible NMDAR antagonists MK 801, with $K_i < 1.0$.[58] Effects of natural EPF on neuronal functional activity confirmed the reversible blocking effects of peptide in LPed8 neuron isolated from *Planorbarius corneus* mollusk (see Figure 5.7).[59] The application of Fischer's method for analyses of CNS glutamate receptor binding sites allowed the theoretical structure of potential EPFs to be calculated.[60]

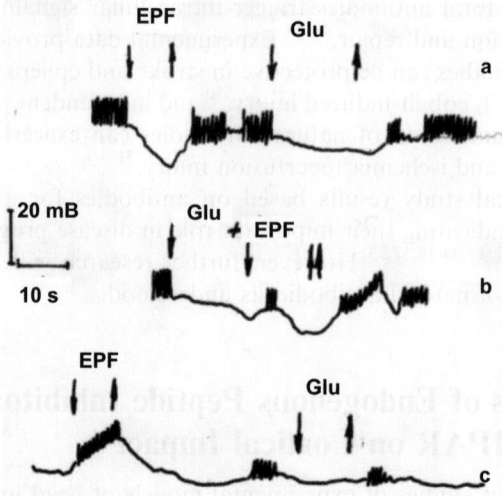

Figure 5.7   Modification of neuronal response to L-glutamate by natural endogenous
peptide factor (EPF). a: the 30 s application of EPF (10 µg) to neuronal
membrane of LPed8 neuron isolated from *Planorbarius corneus* mollusk
evoked the hyperpolarization potential of 20 ± 5 mV. After 30 s washing,
the application of glutamate (1×10^{-3} M) resulted in hyperpolarization of
10 ± 3 mV with hyperpolarization shift of 3–5 mV and 25–30 s desensiti-
zation. b: the combined application of glutamate and EPF quenched
spontaneous activity of neuron. c: short-term application of EPF on
neuronal membrane reversibly reduced glutamate-evoked potential
(10 ± 3 mV) for about 15 s.

5.4.2   Pharmacological Properties and Functional
Activity of Synthetic Peptides

Glutamic acid and 20 other N- and O-derivatives of aminodicarboxylic acids
have been tested for their ability to inhibit binding of ^{3}H-L-glutamate to cortical synaptic membranes. The presence of two anionic or anionic/cationic
groups was necessary to compete for binding site of glutamate receptor.

The effect of various synthetic double glutamate derivatives on specific
binding activity of L-glutamate to synaptic membranes was tested.[61] The
inhibition efficacy $K_i = 3.5$ of di-[1,3-di(hydroxycarbonyl)-propyl-1] amide of
succinic acid was registered and it was suggested that derivatives of the peptide
may have pharmacological potential in neuroprotection against brain injury
(TBI, and post-traumatic epilepsy). The derivative of double glutamate peptide,
Glyzargin, has been synthesized and tested in preclinical studies.

The exposure of primary cortical neuronal culture and neuroblastoma-
glioma cells to glutamate (1 mM) caused significant cell death, while pretreatment with 10^{-9} M Glyzargin significantly increased neuronal survival
and cell density. The peptide (10^{-8} M) stimulated cell differentiation in

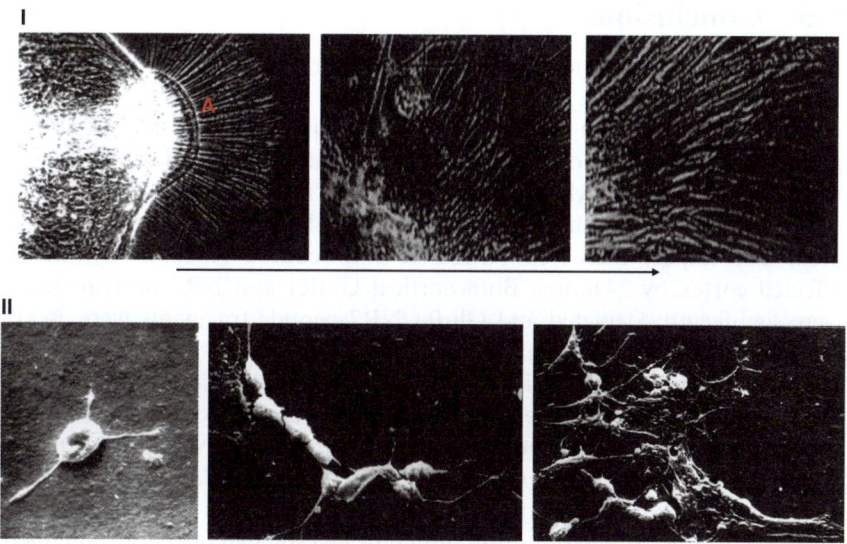

Figure 5.8 I: neurotrophic effect of endogenous peptide factor (EPF) in organotypic tissue. Growth area (A) of cells from rat motor cortex after 1 week of treatment; 3 weeks and 4 weeks showed in magnification ×1300. II: electronic micrographs of spinal motor neuron proliferation after 1, 3, and 4 weeks of treatment showed in magnification of ×400. Arrows show the direction of time progression.

neuronal culture, which suggested that it possesses *in vitro* neuroprotective effect (see Figure 5.8).

5.4.3 Preclinical Studies of Glyzargin Effects

Previous studies have examined the effects of Glyzargin in rats with a low capacity to learn a conditioned food-procuring behavior.[62] It was shown that these rats had a reduced number of post-synaptic glutamate receptors in the striatum and cortex compared to animals with better ability to learn. After intraventricular injections of Glyzargin (0.1–0.3 µg/kg body weight) the learning ability in the first group increased up to 28–30%.

To study the neuroprotective effects of Glyzargin, the peptide was injected intraventricularly (1 µg in 5 µl or 0.3 µM) in a single dose to rats at 3 hours ($n = 6$) after mild CCI. Early treatment with Glyzargin reduced GluR1 peptide levels by 40–45% by 6 hours after experimental concussion. Glyzargin did not alter cerebral blood flow or physiological parameters; however, GluR1 antibody levels down-regulated by 50% from day 3 after mild CCI. GluR1 antibody levels close to that for sham-operated placebo rats by month 2 of the study were detected. No abnormal EEG activity or convulsions were recorded in any of the experimental rats. No significant effect of Glyzargin on NR2 peptide and antibody levels was registered.

5.5   Conclusions

The effects of experimental mild brain injury on alteration of AMPAR and NMDAR subunit expressions, densities in the brain, and resultant degradation products in the blood to trace the time course of AMPA and NR2 peptides and their antibodies trafficking from the CNS have demonstrated that rats with mild CCI, while behaviorally and neurologically normal, had a compromised BBB, with changes in neuronal shapes in the impacted cortical area. The impact increased NMDAR and reduced AMPAR syntheses in affected cortex by 24 hours. Both cortical GluR1 and NR2 protein densities were significantly depleted and GluR1/NR2 peptide fragments were escaping into the bloodstream through the compromised BBB.

In the plasma of rats with mild CCI, GluR1 peptide levels peaked by 3–4 times within 6 hours after the impact; slightly elevated concentrations of GluR1 peptide remained up to 3 months. Significant amounts of NR2 peptide in blood from rats with mild CCI were delayed by 24 hours and their presence may be explained by development of secondary ischemic events around the area of impact. However, this assumption should be investigated further in conjunction with high-resolution MRI scans.

If GluR1 peptide and antibodies are potential biomarkers of mild TBI, treatment with the reversible endogenous peptide inhibitor Glyzargin early after mild brain injury might prevent neurotoxicity and avoid the risk of development of abnormal spiking activity. A single injection of Glyzargin 3 hours after mild CCI significantly down-regulated GluR1 antibodies, and no abnormal EEG activity and convulsions were recorded in all experimental rats after 90 days.

Although this preliminary study was limited to a small group of animals with mild CCI treated using a synthetic peptide derived from a natural peptide inhibitor of AMPAR, Glyzargin showed promising results for preventing sequelae after experimental concussion. Larger scale experimental studies with various models of TBI models (e.g. mild-moderate severity, repetitive cerebral concussions, contusion) should be performed to obtain more data on potential applications of Glyzargin.

Acknowledgments

The authors would like to express gratitude to Dr S. Hoffman (Emory University, Atlanta, GA, USA), Dr J. B. Justice (Emory University, Atlanta, GA, USA), A. Tcherepanov (SUNY, New York, NY), and Dr O. Granstrem (Geropharm, St. Petersburg, Russia) for collaboration in experimental research.

References

1. D. Pohl, M. J. Ishmaru, P. Bittigau, D. Stadhaus, C. Hubner, J. W. Olney, L. Turski and C. Ikonomidou, *Proc. Natl. Acad. Sci. USA*, 1999, **96**, 2508.

2. C. Ikonomidou, V. Stefovska and L. Turski, *Proc. Natl. Acad. Sci. USA*, 2000, **97**, 12885.
3. A. Biegon, P. A. Fry, C. M. Paden, A. Alexandrovich, J. Tsenter and E. Shohami, *Proc. Natl. Acad. Sci. USA*, 2004, **101**, 5117.
4. J. Schumann, G. A. Alexandrovich, A. Biegon and R. Yaka, *J. Neurotrauma*, 2008, **25**, 945.
5. B. C. Albensi, P. G. Sullivan, M. B. Thompson, S. W. Scheff and M. P. Mattson, *Exp. Neurol.*, 2000, **162**, 385.
6. M. J. Sanders, T. J. Sick, M. A. Perez-Pinzon, W. D. Dietrich and E. J. Green, *Brain Res.*, 2000, **861**, 69.
7. J. H. Yi and A. S. Hazell, *Neurochem. Int.*, 2006, **48**, 394.
8. M. Arundine and M. Tymianski, *Cell Mol. Life Sci.*, 2004, **61**, 657.
9. R. Z. Han, J. J. Hu, Y. C. Weng, D. F. Li and Y. Huang, *Neurosci. Bull.*, 2009, **25**(6), 367.
10. T. Costa, L. C. Constantino, B. P. Mendonca, J. G. Pereira, B. Herculano, C. I. Tasca and C. R. Boeck, *J. Neurosci. Res.*, 2010, **88**(6), 1329.
11. M. U. Gappoeva, G. A. Izykenova, O. K. Granstrem and S. A. Dambinova, *Biokhimia*, 2003, **68**, 696.
12. S. A. Dambinova, *Glutamate Neuroreceptors*, Nauka, Leningrad, 1988 (in Russian).
13. R. E. Twyman, L. C. Gahring, J. Spiess and S. W. Rogers, *Neuron*, 1995, **14**(4), 755.
14. M. J. During, C. W. Symes, P. A. Lawlor, J. Lin, J. Dunning, H. L. Fitzsimons, D. Poulsen, P. Leon, R. Xu, B. L. Dicker, J. Lipski and D. Young, *Science*, 2000, **287**, 1453.
15. G. E. Hardingham, *Biochem. Soc. Trans.*, 2009, **37**, 1147.
16. K. Beauchamp, H. Mutlak, W. R. Smith, E. Shohami and P. F. Stahel, *Mol. Med.*, 2008, **14**(11–12), 731.
17. A. I. Gorodinskii and S. A. Dambinova, *Exp. Biol. Med.*, 1986, **102**, 1195 (in Russian).
18. S. A. Dambinova, *Clin. Lab. Int.*, 2008, **32**, 7.
19. S. A. Dambinova, *IVD Technol.*, 2007, **3**, 15.
20. A. Pitkanen and T. K. McIntosh, *J. Neurotrauma*, 2006, **23**(2), 241.
21. Y. P. Tang, Y. Noda, T. Hasegawa and T. Nabeshima, *J. Neurotrauma*, 1997, **14**, 851.
22. O. Zohar, S. Schreiber, V. Getslev, J. P. Schwartz, P. G. Mullins and C. G. Pick, *Neuroscience*, 2003, **118**, 949.
23. N. Henninger, S. Dützmann, K. M. Sicard, R. Kollmar, J. Bardutzky and S. Schwab, *Exp. Neurol.*, 2005, **195**(2), 447.
24. R. Morris, *J. Neurosci. Methods*, 1984, **11**, 47.
25. J. B. Bederson, L. H. Pitts, M. Tsuji, M. C. Nishimura, R. L. Davis and H. Bartkowski, *Stroke*, 1986, **17**, 472.
26. Y. P. Tang, Y. Noda, T. Hasegawa and T. Nabeshima, *J. Neurotrauma*, 1997, **14**, 863.
27. M. S. Beattie, A. R. Ferguson and J. C. Bresnahan, *Eur. J. Neurosci.*, 2010, **32**(2), 290.

28. S. Taliani, I. Pugliesi and F. Da Settimo, *Curr. Top. Med. Chem.*, 2011, **11**(7), 860.

29. U. Wullner, D. G. Standaert, C. M. Tesla, G. B. Landwehrmeyer, M. V. Catania, J. B. Penney, Jr. and A. B. Young, *Brain Res.*, 1994, **647**(2), 209.

30. Z. Fei, X. Zhang, H.-M. Bai, X.-F. Jiang, X. Li, W. Zhang and W. Hu, *BMC Neurosci.*, 2007, **8**, 96.

31. H. B. Pollard, M. Srivastava, O. Eidelman, C. Jozwik, S. W. Rothwell, G. P. Mueller, D. M. Jacobowitz, T. Darling, W. B. Guggina, J. Wright, P. L. Zeitlin and C. P. Paweletz, *Proteomics Clin. Appl.*, 2007, **1**(9), 934.

32. J. Hanrieder, M. Wetterhall, P. Enblad, L. Hillered and J. Bergquist, *J. Neurosci. Methods*, 2009, **177**(2), 469.

33. M. O. Sjödin, J. Bergquist and M. Wetterhall, *J. Chromatogr. B. Analyt. Technol. Biomed. Life Sci.*, 2010, **878**(22), 2003.

34. S. A. Dambinova, G. A. Khounteev, G. A. Izykenova, I. G. Zavolokov, A. Y. Ilyukhina and A. A. Skoromets, *Clin. Chem.*, 2003, **49**, 1752.

35. K. J. Becker, *Neuroscience*, 2009, **158**(3), 1090.

36. R. Siman, T. K. McIntosh, K. M. Soltesz, Z. Chen, R. W. Neumar and V. L. Roberts, *Neurobiol. Dis.*, 2004, **16**(2), 311.

37. K. K. Jain, *The Handbook of Biomarkers*, Humana Press, New York, 2010, p. 382.

38. P. I. Andrews and J. O. McNamara, *Curr. Opin. Neurobiol.*, 1996, **6**(5), 673.

39. A. Shikuev, A. Bagumyan, U. Danilenko and S. Dambinova, *J. Neurotrauma*, 2011, **28**, A-55.

40. D. McGonagle and M. F. McDermott, *PLoS. Med.*, 2006, **3**(8), e297.

41. Y. Zhang and P. Popovich, *Discov. Med.*, 2011, **11**(60), 395.

42. J. J. Archelos and H. P. Hartung, *Trends Neurosci.*, 2000, **23**(7), 317.

43. J. P. De Rivero Vaccari, G. Lotocki, A. E. Marcillo, W. D. Dietrich and R. W. Keane, *J. Neurosci.*, 2008, **28**(13), 3404.

44. B. Diamond, P. T. Huerta, P. Mina-Osorio, C. Kowal and B. T. Volpe, *Nat. Rev. Immunol.*, 2009, **9**(6), 449.

45. P. G. Popovich and E. E. Longbrake, *Nat. Rev. Neurosci.*, 2008, **9**(6), 481.

46. A. Trivedi, A. D. Olivas and L. J. Noble-Haeusslein, *Clin. Neurosci. Res.*, 2006, **6**(5), 283.

47. M. R. Ehrenstein and C. A. Notley, *Nat. Rev. Immunol.*, 2010, **10**(11), 778.

48. B. R. Wright, A. E. Warrington, D. D. Edberg and M. Rodriguez, *Arch. Neurol.*, 2009, **66**(12), 1456.

49. J. Watzlawik, E. Holicky, D. D. Edberg, D. L. Marks, A. E. Warrington, B. R. Wright, R. E. Pagano and M. Rodriguez, *Glia*, 2010, **58**(15), 1782.

50. S. A. Dambinova, O. K. Granstrem, A. Tourov, R. Salluzzo, F. Castello and G. A. Izykenova, *J. Neurochem.*, 1998, **71**, 2088.

51. M. Zhang, W. G. Austen, Jr., I. Chiu, E. M. Alicot, R. Hung, M. Ma, N. Verna, M. Xu, H. B. Hechtman, F. D. Moore, Jr. and M. C. Carroll, *Proc. Natl. Acad. Sci. USA*, 2004, **101**(11), 3886.

52. S. A. Dambinova, G. A. Izykenova, S. V. Burov, E. V. Grigorenko and S. A. Gromov, *J. Neurol. Sci.*, 1997, **152**(1), 93.

53. A. Vincent, S. R. Irani and B. Lang, *Curr. Opin. Neurol.*, 2010, **23**(2), 144.
54. E. I. Gusev, V. I. Skvortsova, S. A. Dambinova, K. S. Raevskiy, A. A. Alekseev, V. G. Bashkatova, A. V. Kovalenko, V. S. Kudrin and E. V. Yakovleva, *Cerebrovasc. Dis.*, 2000, **10**, 49.
55. W. Chen, Q. Ma, H. Suzuki, R. Hartman, J. Tang and J. H. Zhang, *Stroke*, 2011, **42**(3), 764.
56. A. A. Skoromets, L. V. Stakhovskaia, A. A. Belkin, K. V. Shekhovtsova, O. B. Kerbikov, D. V. Burenchev, O. V. Gavrilova and V. I. Skvortsova, *Zh. Nevrol. Psikhiatr. Im S.S. Korsakova*, 2008, **22**(Suppl 32) (in Russian).
57. S. A. Dambinova and A. I. Gorodinskii, *Biokhimia*, 1984, **49**, 67 (in Russian).
58. I. Talukder, P. Borker and L. P. Wolmuth, *J. Neurosci.*, 2010, **30**(35), 11792.
59. M. N. Margulis, S. A. Gapon, N. I. Kudriashova and S. A. Dambinova, *Neurophysiology*, 1988, **20**, 137 (in Russian).
60. S. A. Dambinova and N. Chromov-Borisov, *Proc. Acad. Sci. USSR*, 1984, **274**, 467 (in Russian).
61. L. B. Piotrovskii, M. N. Dumpis, V. D. Ioffe, A. I. Gorodinskii and S. A. Dambinova, *Proc. Acad. Sci. USSR*, 1986, **286**, 235 (in Russian).
62. I. V. Karpova, A. I. Gorodinsky, N. F. Suvorov and S. A. Dambinova, *I.M. Sechenov' Physiol. J.*, 1994, **80**, 72 (in Russian).

CHAPTER 6
Mitochondrial Dysfunctions and Markers of Spinal Cord Injury

ALEXANDER V. PANOV

WellStar College of Health & Human Services Kennesaw, Kennesaw State University, Laboratory of Brain Biomarkers, 1000 Chastain Road, Bldg. 41, Kennesaw, Georgia 30144, USA
Email: apanov@kennesaw.edu

6.1 Introduction

Traumatic brain injury (TBI) and spinal cord injury (SCI) represent a great socioeconomic impact on any society because of their high mortality and in that they disable the most active members of the population. In the USA, yearly incidence of TBI is between 1800 and 2500 per million;[1] incidence of SCI is lower, approximately 700 cases per million.[2] A modest estimate of total lifetime costs per TBI case in the USA is around $200 000,[3] whereas for patients with SCI, it is close to $1 500 000.[4]

In patients with spinal cord injury, the primary or mechanical trauma seldom causes total transection, even though functional loss may be complete.[4,5] The primary injury is immediately followed by various systemic and local pathological events, known as the secondary injury mechanisms, that within a few hours may turn an incomplete injury into a complete injury of the traumatized spinal cord. Therefore, the quality of the first aid and correct handling of traumatized persons is critical. In Canada, a significant reduction in the proportion of patients with complete SCI, from approximately 65% to approximately 45%, was achieved mainly due to improved management, including better first aid, evacuation, and transportation, and improved

RSC Drug Discovery Series No. 24
Biomarkers for Traumatic Brain Injury
Edited by Svetlana A. Dambinova, Ronald L. Hayes and Kevin K. W. Wang
© Royal Society of Chemistry 2012
Published by the Royal Society of Chemistry, www.rsc.org

treatment in specialized regional units delivering the first aid.[5,6] With the possible exception of methylprednisolone,[7] there is no effective means of treatment of acute SCI.[4] However, in recent years, use of methylprednisolone for the treatment of acute SCI has been disappointing,[8] contradicting results obtained in animal models of SCI.[9] Therefore, the search for pharmacological treatment of acute SCI to prevent secondary injury mechanisms, together with further improvements of the technical methods of fixation and evacuation of patients, are of paramount importance.

Lack of reliable pharmacological methods to fight secondary injury mechanisms, including contradicting results with the use of methylprednisolone, have been, to a large degree, due to limited knowledge of the specifics of spinal cord energy metabolism, physiology, and pathology of myelin. Only recently has it been shown that there are significant differences between brain and spinal cord mitochondria that might explain different consequences for TBI and SCI.[10,11] After TBI, pathological events, including apoptotic death of the brain's neurons, usually develop relatively slowly, whereas after SCI, neurons die, apparently by necrosis, within few hours after trauma.

A large number of publications on SCI is devoted to the search for biological markers that would help clinicians predict outcome of SCI and follow-up dynamics of treatment. Recently, several reviews and papers on various markers of SCI have been published.[12–24] The purpose of this chapter is to evaluate current knowledge in the field of SCI markers from the point of recently acquired knowledge on the role of mitochondria in pathogenesis of neurodegenerative diseases and the specifics of mitochondrial energy metabolism in the spinal cord. Because the most critical events for SCI outcome occur during the first 24 hours, this chapter focuses on the role of mitochondria in the death of neurons subjected to secondary injury mechanisms. This knowledge will help in finding specific markers that would characterize the severity of secondary injury mechanisms and thus predict perspectives for preservation of neuronal functions.

6.2 Brain and Spinal Cord Mitochondria Respond Differently to Trauma

After SCI, even in the absence of physical disruption of the spinal cord, death of neurons may occur within the first 6–12 hours due to development of secondary injury mechanisms, which involve mitochondrial dysfunction, whereas after TBI, neurodegeneration occurs much more slowly. To a large extent, differences between brain and spinal cord in responses to trauma can be explained by differences in mitochondrial metabolism.

6.2.1 Metabolic Differences between Spinal Cord and Brain Mitochondria

There is strong evidence that during neuronal activation, postsynaptic mitochondria are exposed to increased levels of pyruvate and glutamate,[25,26] and simultaneous oxidation of pyruvate and glutamate may significantly increase rates of ATP synthesis and state 4 generation of reactive oxygen species (ROS). This is because brain and spinal cord mitochondria have high activities of mitochondrial transaminases that convert glutamate and pyruvate to α-ketoglutarate, which is further oxidized to succinate. This metabolic feature is specific only for brain mitochondria.[27] Qualitatively, brain and spinal cord mitochondria have similar responses to the above substrate mixtures; however, they also have important distinctions in their major functions.

Panov *et al.*[10] have shown that with "classical" substrates; that is, with glutamate or pyruvate in the presence of malate, spinal cord mitochondria have 30% to 40% lower respiratory activities during oxidative phosphorylation (state 3) and uncoupling (state 3U), results that were in concordance with those presented earlier by Dave *et al.*[28] and Sullivan *et al.*[29] Results of Western blot tests suggested that lower respiratory activity of spinal cord mitochondria was more likely associated with lower contents of the enzyme complexes responsible for electron transport and ATP synthesis. Forner *et al.*[30] presented a quantitative proteomic comparison of rat mitochondria from muscle, heart, and liver and concluded that about 40% of the 689 proteins identified can be associated with one tissue, either because of tissue-unique expression or differential expression. Thus, substantial variations in mitochondrial proteins and functions in various tissues reflect close correlations between tissue and mitochondrial function.

However, when brain and spinal cord mitochondria oxidized physiologically relevant substrate mixtures, producing significant amounts of succinate,[27] increases in respiratory rates were relatively higher with spinal cord mitochondria, which may compensate somewhat for the lower contents of respiratory enzymes in ATP production. Because respiration activation with glutamate plus pyruvate plus malate was bound with formation of succinate within the tricarboxylic citric acid (TCA) cycle, it was sensitive to inhibition by oxaloacetate and malonate, strong inhibitors of succinate dehydrogenase (SDH).

6.2.2 Relationships between Respiration and the Mitochondrial Calcium Retention Capacity

There is strong evidence that Ca^{2+}-induced mitochondrial permeability transition (mPT) is one of the major mechanisms that induce apoptosis.[31,32] Permeability transition is a sudden increase in conductance of the mitochondrial inner membrane to solutes up to 12 000 Da due to opening of the Ca^{2+}-dependent pore. The amounts of calcium phosphate that mitochondria can sequester before opening the permeability transition pore strongly depend on mitochondrial respiratory activity, which is reflected in the H^+/Ca^{2+} ratio.[33,34] Even small differences in the H^+/Ca^{2+} ratios may result in large dissimilarities in the mitochondrial capacity to sequester CaPi.[34,35] Spinal cord mitochondria have significantly lower H^+/Ca^{2+} ratios compared with brain mitochondria.[10] Correspondingly, calcium retention capacity (CRC) values for spinal cord mitochondria were 40% to 50% lower than those for brain mitochondria.[10]

Following trauma, there is a significant release of Ca^{2+} from cellular stores, which potentially may cause mPT and initiate apoptotic or necrotic cell death.[39] Of the total Ca^{2+} content in parenchymal tissues, only a small fraction exists in a free form, and normal concentrations of $[Ca^{2+}]_{Free}$ are below 1 μM.[36] Figure 6.1 shows that CRC of the non-synaptic brain mitochondria, isolated from 1 g of tissue, exceeded total Ca^{2+} content in tissue at least twofold. This makes it unlikely that Ca^{2+}-induced permeability transition, and hence excitotoxic cell death, occur in the brain in the absence of additional pathogenic factors. Support for the glutamate-induced Ca^{2+}-excitotoxic cell death hypothesis comes mainly from experiments with cultured neurons.[32,37] Cell culture media usually contain at least 0.5 mM of Ca^{2+},[38] which is more than enough to cause mPT followed by excitotoxic cell death.

Figure 6.1 also shows that total Ca^{2+} content in the spinal cord tissue is 8 times higher than in the brain. When normalized for 1 g of tissue, spinal cord mitochondria could sequester only 10% to 20% of the total Ca^{2+} available in the spinal cord. This suggests that in the spinal cord, most of the tissue Ca^{2+} is located in myelin.[10] It was shown that following trauma, ample amounts of Ca^{2+} are released, more likely due to damages to myelin.[39] Halestrap[40] suggested that opening of mPTP initiated by Ca^{2+} usually triggers necrotic cell death.

6.2.3 Distinctive Properties of ROS Generation in Brain and Spinal Cord Mitochondria

In neuronal tissue, mitochondria are the major source of reactive oxygen species (ROS) production,[41] which is an indispensable feature of aerobic metabolism. It is generally accepted that during oxidation of NAD-dependent substrates, the limiting step is at the NADH-dehydrogenase site of Complex I.[27,35] With succinate as a substrate, production of ROS is significantly higher, because Complex II, which is also part of the TCA cycle–SDH, feeds electrons into the mitochondrial pool of CoQ_{10} (ubiquinone, Q). The reduced ubiquinone (QH_2) in turn reduces sites on Complex I that can generate superoxide radicals at a high rate. This process, known as reverse electron transport, is energy dependent and inhibited by rotenone. Some researchers deny the significance of succinate in ROS production on the pretext that succinate concentration is too low in mitochondria.[41–43] As shown recently, brain and spinal cord mitochondria may produce succinate in the presence of pyruvate plus malate, and even with glutamate plus malate, a large part of ROS production may be associated with oxidation of succinate.[27,44] A particularly large production of succinate-dependent ROS was observed in the presence of glutamate plus pyruvate plus malate.[10,11,27] With these substrates, increased succinate oxidation was caused by a dramatic increase in mitochondrial α-ketoglutarate due to high activities of aminotransferase.[45,46] In addition, γ-aminobutyric acid (GABA) is catabolized in postsynaptic mitochondria with formation of succinate.[47] Thus, in the excited neurons, succinate is an indispensable mitochondrial metabolite. Importantly, ROS generation associated with the succinate-dependent reverse electron transport is a subject for phenotypic variations.[44]

One of the other objections to the importance of reverse electron transport (RET) in ROS generation argues that, because RET is energy dependent, in the functioning cell diminished mitochondrial energization will inhibit production of ROS.[42] The latter objection is valid for the most perpetually functioning organs, such as heart, kidney, and liver, but it is only partially applicable to the brain and spinal cord, where most of the mitochondria are located at axonal and dendritic junctions.[48] At the narrow spaces of synaptic junctions, there is no other task for mitochondria beside provision of ATP for restoration of ionic composition in excited synapses. If the neurons are not excited, mitochondria become fully energized and produce ROS at a high rate. Therefore, axonal and synaptic junctions, including neuromuscular junctions, are particularly vulnerable to oxidative damage if neurons are not excited for some reason or during hypoxia, a common complication after SCI and TBI.

The following metabolic scenario for the neuronal mitochondria located at synaptic junctions is suggested. During neuronal activity, postsynaptic mitochondria are exposed to increased levels of pyruvate and glutamate, which enhance ATP production due to specific interactions between the glutamate transforming and the TCA cycle enzymes. This effect of the substrate mixtures is specific for brain and spinal cord mitochondria.[10,27] As soon as activation of a synapse ceases, transport of glutamate from the synaptic cleft also stops. As a result, activation of respiration induced by the simultaneous presence of glutamate and pyruvate will also cease, because glutamate becomes exhausted. Thus, the neuromediator glutamate controls energy metabolism in the brain not only at the level of astrocytes by enhancing production of lactate, but also at the level of neurons by controlling the activity of malate aspartate shuttle (MAS) and specific interactions between aminotransferases.

In the absence of neuronal activation, SDH (Complex II) in the postsynaptic mitochondria is inhibited by oxaloacetate (OAA).[10] The physiological significance of this inhibition is to prevent excessive ROS production associated with the reverse electron transport. This is particularly important for neurons because with pyruvate, which is the major mitochondrial substrate in resting neurons,[25,49] there is a substantial production of succinate, and thus a possibility of increased RET-dependent ROS production.[10,11,27] During increased neuronal activity, glutamate and pyruvate temporarily release inhibition of SDH. Increased ATP production prevents generation of ROS.[42,50] In general, spinal cord mitochondria oxidize succinate at higher rates than brain mitochondria, and have lower intrinsic inhibition of SDH.[10,11] As a result, spinal cord mitochondria are more vulnerable to oxidative stress than brain mitochondria.[10,11]

6.2.4 Activation of Glutamate Receptors Stimulates Acute Excitotoxic Death of Neurons and Oligodendrocytes in Spinal Cord Injury

Both brain and spinal cord trauma result in excessive release of the neuromediator glutamate, which is known to stimulate excitotoxic death of neurons.[39,51]

The excitotoxicity of glutamate is closely associated with the uncontrolled entry of Ca^{2+} into neurons through glutamate receptor-controlled calcium channels and initiation of the Ca^{2+}-induced mitochondrial permeability transition that is the final deadly step in the sequence of events initiated by increased glutamate. However, significant differences between the brain and spinal cord in responses to increased glutamate due to different distribution of glutamate receptors have been found.

Glutamate receptors are divided into ionotropic types, which include *N*-methyl-D-aspartate (NMDA), 2-amino-3-(5-methyl-3-oxo-1,2,-oxazol-4-yl)-propanoic acid (AMPA)/kainate receptors, and metabotropic glutamate receptors that operate through G-protein membrane receptors.[52] NMDA and AMPA/kainate receptors are also found in the non-neuronal cells, such as glial cells and oligodendrocytes of white matter.[39,53,54] Therefore, glutamate-induced excitotoxicity is responsible for the death of both neurons and glia.[51] Comparative studies of murine spinal cord and cortical cultures have shown that spinal cord neurons were much less vulnerable to activation of NMDA and more vulnerable to activation of AMPA and kainate receptors.[55] Experiments on animals with incomplete contusive SCI have also shown that a large and functionally important proportion of the tissue loss was caused by secondary injury mediated by local AMPA/kainate receptors.[56]

Recently, it was established that in spinal cord injury, both NMDA and AMPA/kainate receptors play important roles not only in glutamate toxicity-induced death of neurons but also in post-traumatic white matter degeneration.[51,57,58] Early and progressive demyelination is an important feature of the secondary injury of traumatized spinal cord.[59] In the brain, white and gray matter comprise equal proportions, while in the spinal cord, white matter greatly predominates.[60] We suggest that demyelination releases large amount of calcium ions, which hold together the sheets of myelin, and thus glutamate neurotoxicity in traumatized spinal cord causes mitochondrial mPT and necrotic death of neurons. In the brain, the calcium retention capacity (CRC) of mitochondria exceeds the total tissue calcium content at least two fold,[10,11,27] therefore diminishing the possibility of excitotoxic death of neurons in the presence of excessive glutamate. In the spinal cord, the CRC value for spinal cord mitochondria is 40% to 50% lower than in brain mitochondria, and the total tissue Ca^{2+} content is eightfold higher than in the brain.[10]

The important role of excitotoxicity in the pathogenesis of secondary injury led to a number of attempts to prevent complications for neurons and white matter with antagonists of NMDA and non-NMDA receptors.[51,54,56,61] A comprehensive description of markers for NMDA and AMPA/kainate receptors can be found in this book (see Chapter 4).

6.3 The Concept of Primary and Secondary Acute Spinal Cord Injury

Managing patients with SCI during the first hours after trauma is often a great challenge because the trauma may involve more than the spinal cord.

Besides, patients may be intoxicated by alcohol or drugs, and evacuation may be difficult.[5,6] In most cases, it is impossible to conduct a thorough examination and take samples for laboratory tests. Therefore, as with all other neurodegenerative diseases, animal models are most important for understanding the pathological events occurring after SCI and TBI.

The time line of events developing after SCI can be briefly presented as follows:

- time 0: the moment of injury
- few minutes to 24 hours or longer: the secondary injury events begin: hemorrhage, ischemia, drop of blood pressure, edema, inflammation, disruption of blood-brain barrier, local release of neuromediators, calcium ionic disbalance, oxidative stress, acute demyelination
- 2–8 hours: death of local neurons
- 1 day and later: some symptoms of the secondary injury mechanisms continue (e.g. inflammation, low blood pressure, edema), the beginning of systemic metabolic events caused by neuronal dysfunctions: reduction in general metabolic activity and negative nitrogen balance associated with degradation of the spinal cord tissue and skeletal muscle below the site of the injury
- 7–14 days or later: period of gradual stabilization.[4,39,62–69]

Runge *et al.*[70] studied temporal evolution of acute spinal cord injury in rats using magnetic resonance imaging (MRI). They showed that the intensity of cord enhancement in the region of injury, after intravenous contrast injection, was at a maximum on the day of injury, decreased in a steady fashion thereafter, and disappeared on day 28. This study clearly showed that the blood-brain barrier was severely compromised on the day of injury and gradually restored its function in the course of 4 weeks. However, both MRI and computerized tomography (CT) do not always reflect the severity of the functional and biochemical events that occur after trauma. A comparative study of CT and blood samples for the biochemical markers of TBI have shown that some patients without the CT-confirmed intracerebral pathology had significantly increased levels of blood serum S-100β protein and the neuron-specific enolase (NSE), markers indicating neuronal injury.[12] Numerous studies have established that S-100β protein and NSE are to date the most reliable markers of SCI[19–24,62,71] and TBI[72–74] that can be correlated with trauma severity. However, in cases of SCI, the time course of these markers is different from that in TBI.

6.4 Early Markers of Spinal Cord Injury

In one of the latest papers published on SCI markers, Lubieniecka *et al.*[71] reported the results of a typical broad-scale search for SCI biomarkers using a quantitative liquid chromatography–mass spectrometry analyses of cerebrospinal fluid (CSF) collected from rats 24 hours after either a moderate or severe SCI. The authors identified a panel of 42 putative biomarkers of SCI,

10 of which represent potential biomarkers of SCI severity. Several of the potential biomarkers are immune response proteins, including B2M, Serpinc1, A2M, C3, ApoA1, ApoH, and Ambp. It is unclear, however, how these and many others biomarkers are specific for the actual mechanism of neurodegeneration at the site of injury. Twenty-four hours after SCI a number of events, both pathological and adaptive, occur simultaneously; therefore, such markers may have little utility for clinicians. In addition, immunological markers may reflect more the severity of inflammation and only indirectly be related to severity of the trauma. The irrelevance of the markers found in CSF 24 hours after SCI was also confirmed by the fact that the authors did not find any alterations in the relative abundance of S100β and NSE proteins.

Two recent studies investigated the time course of changes in the contents of S100β and NSE in the blood serum and CSF after acute SCI of different severity.[18,20] Loy *et al.*[18] found that increases in serum levels of NSE were observed for 200-kdyn (3.1-fold) and 150-kdyn (2.3-fold) injury groups at 6 hours after injury, which decreased by 73.7% and 65.2% at 24 hours after SCI, respectively; the levels were still greater than in sham animals. The 200- and 150-kdyn injury groups were not different at either time point. S-100β serum levels increased at 6 hours in the 200-kdyn injury group, and no differences from sham levels were seen at 24 hours. No differences in total protein concentrations were observed between the injury and control groups. In similar experiments, Cao *et al.*[20] compared the dynamics of S-100β and NSE proteins after mild and severe SCI. They found no significant difference in NSE or S-100β protein levels between the control group and SCI groups at 30 minutes after injury. However, a significant increase in NSE and S-100β protein levels in serum and CSF was observed at 2 hours after injury, and they reached maximum levels at 6 hours. At 24 hours after injury, although NSE and S-100β protein levels in the SCI groups were still higher than that in the control group, the graded SCI groups did not differ significantly with respect to either NSE or S-100β protein levels. Over the course of the experiment, changes in the serum S-100β protein level were similar to those that occurred in the CSF. This is understandable, because according to Runge *et al.*[71] the blood-brain barrier was compromised.

In the literature there is much reasoning about what are ideal biomarkers.[20,71] Two biomarkers, S-100β and NSE, correlate reasonably well with trauma severity, change with time, and have been consistently observed by researchers studying SCI[19–24,63] and TBI.[72–74] In the literature on markers, authors usually do not give specifics about their origin and meaning. But without specifics, it is difficult to evaluate the significance and relationships between markers and answer "simple" questions such as "Are these markers reliable?", "What do they mean?", and "How useful are they?". We will try to answer some of these questions.

6.4.1 Neuron-Specific Enolase

Enolase, also known as phosphopyruvate dehydratase, is a metalloenzyme responsible for catalysis of the conversion of 2-phosphoglycerate (2-PG) to phosphoenolpyruvate (PEP) in the forward or catabolic direction in the second half of the glycolytic pathway. In the reverse reaction (anabolic pathway), which occurs during gluconeogenesis, the same enzyme catalyzes hydration of PEP to 2-PGA. The enzyme is a homodimer and has a molecular mass of 82 000–100 000 Da, depending on the isoform, and is present in all tissues and organisms capable of glycolysis. In vertebrates, the enzyme occurs as three isoenzymes: α-enolase is found in a variety of tissues including liver, whereas β-enolase is almost exclusively found in muscle tissues, and γ-enolase, also known as NSE or enolase 2 (ENO2), is found in neurons and neuroendocrine tissues.[75] In humans, NSE is encoded by the ENO2 gene.[75] Although NSE is a glycolytic enzyme, it is astroglia, which have highly active aerobic glycolysis, that provides neurons with lactate.[76] Evidently, in neurons NSE has other functions, similar to α-enolase in the liver. Recent findings have shown that α-enolase performs several important functions in addition to innate glycolytic function and plays an important role in several biological and pathophysiological processes in the liver.[75]

As a marker, release of NSE after acute SCI or TBI into the blood stream and CSF after trauma is a sign of neurodegeneration. Because the increase in this biomarker is relatively rapid and short-termed after SCI, this suggests NSE reflects death of local spinal cord neurons caused by secondary injury mechanisms. In the absence of pathological events, which comprise the secondary injury, neurodegeneration caused by "clean" surgical transection of the spinal cord, known as Wallerian degeneration, develops slowly.[77]

6.4.2 S-100 Protein

S-100 protein is so-called for its solubility in a 100% saturated solution of ammonium sulfate, and was described as an acidic protein unique to the nervous system. S-100 has no enzymatic activity.[78] S-100 was shown to be involved in several activities, including memory processes, regulation of diffusion of monovalent cations across membranes, modulation of the physical state of membranes, regulation of the phosphorylation of several proteins, and control of the assembly-disassembly of microtubules. Some of these effects are strictly Ca^{2+}-dependent, while others are not. The classical S-100, i.e. S-100 purified from bovine brain, has a molecular mass of 21 000 Da and is made of three subunits linked by non-covalent bonds. S-100β binds Ca^{2+} with apparent $K_{d1} = 60 \mu M$ and apparent $K_{d2} = 0.2 mM$ at pH 8.3. There are two forms of S-100 proteins, cytoplasmic and membrane-bound.[77] Cytoplasmic and membrane-bound S-100 from rat brain consists mostly of S-100β.[79] Mata *et al.*[80] have found that S-100β is preferentially distributed in myelin-forming Schwann cells, suggesting that in these cells, S-100β expression may be related to axon diameter and degree of myelination.

Axons are made metabolically efficient by myelination, which enables saltatory conduction, i.e. conduction of nerve impulses, in which the impulse jumps rapidly from one node of Ranvier to the next. In a recent review, Campbell and Mahad[81] stressed the importance of mitochondria for maintaining the structural integrity of myelinated axons, which is illustrated by neuroaxonal degeneration in primary mitochondrial disorders and trauma. Recently, the total calcium content in the spinal cord tissue has been found to be 8 times higher than in brain tissue (see Figure 6.1).[10] Evidently, the majority of this calcium participates in holding together the sheets of myelin surrounding the axon. One of the pivotal functions of mitochondria in a cell is maintenance of calcium homeostasis in the cytoplasm. In the presence of excess of Ca^{2+}, mitochondria open the high conductance pore, which initiates apoptotic or necrotic cell death, so-called excitotoxic cell death.[27,34,35] An important distinction between brain and spinal cord mitochondria has been observed.[10] In the brain, there is not enough tissue Ca^{2+} to cause mitochondrial permeability transition, and thus excitotoxic death of a neuron (see Figure 6.1). In the spinal cord, however, the total amount of Ca^{2+} is 8 times higher than in the brain, which is easily released in trauma or mitochondrial dysfunction.[44] Therefore, the release of approximately 20% of the total spinal cord tissue calcium would be sufficient to initiate mitochondrial permeability transition (see Figure 6.1).[10,11] It is important to note that when mitochondria massively undergo permeability transition, the cells usually die by necrosis, rather than by apoptosis.[39,40,82]

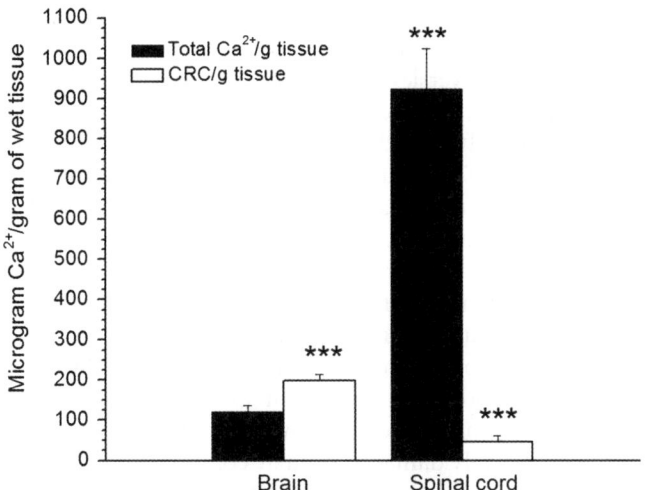

Figure 6.1 The spinal cord and brain tissue total Ca^{2+} contents and the calcium retention capacity of mitochondria isolated from 1 gram of the tissue. Data are expressed as microgram Ca^{2+} per 1 g of wet tissue. Dark bars: total tissue Ca^{2+} content; light bars: mitochondrial calcium retention capacity in the presence oligomycin $2\,\mu g/ml + ADP$ $50\,\mu M$. The data for the spinal cord mitochondria were compared with the corresponding data representing brain mitochondria, and calcium retention capacity (CRC) for mitochondria isolated from 1 g of tissue was compared with the corresponding total tissue calcium. Calcium retention capacity was determined in a medium containing KCl 125 mM, NaCl 10 mM, $MgCl_2$ 0.5 mM, glycyl-glycine 3 mM, pH 7.2, KH_2PO_4 1 mM, and glutamate 5 mM + pyruvate 2.5 mM + malate 2 mM, volume 1 ml. Additions: oligomycin $2\,\mu g/ml$, ADP $50\,\mu M$ (final concentration) were added before mitochondria 0.3 mg/ml, Ca^{2+} 100 nmol/ml. For more details see reference 10.

6.5   Early Involvement of Mitochondria
in Spinal Cord Injury

Neuronal tissue is strictly aerobic and requires large amounts of ATP to fulfill its specific functions. Unlike other tissues, neurons spend relatively little energy for housekeeping, and have low synthetic metabolic activity and regenerative capacity.[83,84] Therefore, all known neurodegenerative diseases involve mitochondrial dysfunctions as obligatory pathogenic mechanisms. Acute SCI is no exception. Sullivan *et al.*[85] studied mitochondria isolated from sections of the traumatized rat spinal cord and found that significant decline in respiratory activity occurred 12 hours after trauma, and further deteriorated at the 24-hour point. The loss of respiratory activity was preceded by progressive increase in oxidative damage of the mitochondria.[85] In a similar set of experiments, Azbil *et al.*[86] studied mitochondrial respiration using synaptosomal preparations. They found that mitochondrial metabolic activity was significantly inhibited as early as 1 hour following SCI. ROS formation and lipid peroxidation were significantly increased, correspondingly, at 4 hours and 1 hour post injury.[86] Ito *et al.*[87] measured the activity of Complex IV of the respiratory chain and found that a 50% inhibition of this terminal respiratory enzyme was evident 15 minutes after SCI. It should be noted that studying mitochondrial functions in the spinal cord is an extremely difficult task. Despite some controversies between the data presented above, it is clear that mitochondria become dysfunctional early after SCI and coincide in time with increased biomarkers S-100β and NSE, which designate damages to myelin and neurons. A large body of literature indicates that oxidative stress increases early after trauma and may precede and facilitate mitochondrial dysfunction.[85,86,88] In rats after experimental SCI, oxidative stress was shown to begin within a few hours after trauma and persisted for 5 days. Methylprednisolone reduced lipid peroxidation within the first 12 hours.[88] Some authors found increased oxidative stress at 5 minutes, 1, and 3 hours, but not at 5 hours post injury.[89]

Therefore, increased oxidative stress occurs early after SCI and may accelerate the destructive processes in myelin and mitochondria. However, the information regarding oxidative stress was obtained indirectly by measuring markers of the oxidatively damaged cellular components such as nitrotyrosine, and protein carbonyls.[85,89] Only recently have direct measurements of ROS production by brain and spinal cord mitochondria been studied.[10,11] It was shown that ROS production by brain and spinal cord mitochondria strongly depends on the types of substrates and thus on the metabolic phenotype of experimental animals.[10,11,44] In brain and spinal cord mitochondria, the major source of ROS is associated with oxidation of succinate, and spinal cord mitochondria produce significantly more ROS than brain mitochondria.[10,11] Unlike mitochondria from other tissues, brain and spinal cord mitochondria may form and oxidize succinate even with pyruvate or glutamate as a substrate.[10,44] In addition, catabolism of γ-aminobutyric acid (GABA) in postsynaptic mitochondria also produces succinate.[47] Based on these data, it is speculated that since methylprednisolone is a highly hydrophobic substance, for clinical use it is injected as a complex with sodium succinate.[14] Therefore, it is possible that in some patients, the injected succinate could facilitate oxidative stress in spinal cord mitochondria and thus oppose the positive effects of the steroid. The effects of succinate on ROS production would strongly depend on the metabolic phenotype of an individual or experimental animal.[12]

Thus, following a SCI, a combination of pathological events, which include damages to myelin accompanied by release of large amount of calcium ions and increased oxidative stress, all result in early damage of mitochondrial integrity, which lead to necrotic or apoptotic death of neurons. Whether neurons die by necrotic or apoptotic pathways depends on the degree of oxidative stress and amount of calcium released.[27,90,91] Other events, when added to secondary injury mechanisms, such as hypoxia and inflammation, also facilitate neuronal death.[4,24]

6.6 Conclusions

Development of pathological and clinically relevant events after SCI can be roughly split into three periods: (i) the moment of SCI and development of acute secondary injury mechanisms, which lead to death of spinal cord neurons within the first 6 to 12 hours after trauma; (ii) time after 12 hours post trauma to several days thereafter, when the secondary injury mechanism continues to exist and the systemic responses develop as a reaction to changed physical and metabolic state of the organism; and (iii) all systems in the organism come to a new steady state. Analysis of literature on biomarkers of SCI have shown that two markers, S-100β protein and NSE, correlate well with severity of SCI and mark the time of myelin disintegration and death of neurons at the trauma site. However, these markers are informative only during the first period following trauma, when the patient may not yet be hospitalized. Therefore, clinical utility of these markers is limited. However, results of investigations of animal models of SCI clearly show that neurons die via the mitochondria-dependent pathway, which involves NMDA and non-NMDA receptors. The literature also shows that NMDA and AMPA/kainite receptors as well as mitochondria are important targets for pharmacological interventions at the early stage of SCI. Thus far, however, little attention has been paid to markers for mitochondria that would reflect the state of these vital organelles in the acute period of SCI. Markers for NMDA and AMPA/kainate receptors have been shown to be of great predictive importance for a number of pathologies.[62] Therefore, using these markers more widely in studies of SCI seems warranted.

Also accentuated is that current understanding of the mechanisms of neuronal death at the acute period of SCI calls for development of methods of pharmacological intervention to prevent secondary injury mechanisms.

References

1. J. Bruns, Jr. and W. A. Hauser, *Epilepsia*, 2003, **44**(Suppl 10), 2.
2. C. B. Harvey, B. B. Rothschild, A. J. Asmann and T. Stripling, *Paraplegia*, 1990, **28**(9), 537.

3. E. Kovesdi, J. Luckl, P. Bukovics, O. Farkas, J. Pal, E. Czeiter, D. Szellar, T. Doczi, S. Komoly and A. Buki, *Acta Neurochir. (Wien)*, 2010, **152**(1), 1.
4. C. H. Tator and M. G. Fehlings, *J. Neurosurg.*, 1991, **75**(1), 15.
5. C. H. Tator, *Br. J. Surg.*, 1990, **77**(5), 485.
6. C. H. Tator, E. G. Duncan, V. E. Edmonds, L. I. Lapczak and D. F. Andrews, *Surg. Neurol.*, **40**(3), 207.
7. M. B. Bracken, *Spine*, 2001, **26**, S47.
8. M. G. Fehlings, *Spine*, 2001, **26**(24 Suppl), S55.
9. E. D. Hall, *J. Neurosurg.*, 1992, **76**(1), 13.
10. A. Panov, N. Kubalik, N. Zinchenko, D. M. Ridings, D. A. Radoff, R. Hemendinger, B. R. Brooks and H. L. Bonkovsky, *Am. J. Physiol. Regul. Integr. Comp. Physiol.*, 2011, **300**(4), R844.
11. A. Panov, N. Kubalik, N. Zinchenko, R. Hemendinger, S. Dikalov and H. L. Bonkovsky, *Neurobiol. Dis.*, 2011, **44**, 53.
12. M. Herrmann, S. Jost, S. Kutz, A. D. Ebert, T. Kratz, M. T. Wunderlich and H. Synowitz, *J. Neurotrauma*, 2000, **17**(2), 113.
13. M. Herrmann, N. Curio, S. Jost, C. Grubich, A. D. Ebert, M. L. Fork and H. Synowitz, *J. Neurol. Neurosurg. Psychiatry*, 2001, **70**(1), 95.
14. J. Ma, L. N. Novikov, K. Karlsson, J. O. Kellerth and M. Wiberg, *Scand. J. Plast. Reconstr. Surg. Hand Surg.*, 2001, **35**(4), 355.
15. S. A. Thomassen, I. L. Johannesen, E. J. Erlandsen, J. Abrahamsen and E. Randers, *Spinal Cord*, 2002, **40**(10), 524.
16. M. Guéz, C. Hildingsson, L. Rosengren, K. Karlsson and G. Toolanen, *J. Neurotrauma*, 2003, **20**(9), 853.
17. N. C. Ringger, B. E. O'Steen, J. G. Brabham, X. Silver, J. Pineda, K. K. Wang, R. L. Hayes and L. Papa, *J. Neurotrauma*, 2004, **21**(10), 1443.
18. D. N. Loy, A. E. Sroufe, J. L. Pelt, D. A. Burke, Q. L. Cao, J. F. Talbott and S. R. Whittemore, *Neurosurgery*, 2005, **56**(2), 391.
19. L. M. Benneker, C. Leitner, L. Martinolli, K. Robert, H. Zimmermann and A. K. Exadaktylos, *Scand. J. Trauma Resusc. Emerg. Med.*, 2008, **16**, 13.
20. F. Cao, X. F. Yang, W. G. Liu, W. W. Hu, G. Li, X. J. Zheng, F. Shen, X. Q. Zhao and S. T. Ly, *J. Clin. Neurosci.*, 2008, **15**(5), 541.
21. M. H. Pouw, A. J. Hosman, J. J. van Middendorp, M. M. Verbeek, P. E. Vos and H. van de Meent, *Spinal Cord*, 2009, **47**(7), 519.
22. E. Kovesdi, J. Luckl, P. Bukovics, O. Farkas, J. Pal, E. Czeiter, D. Szellar, T. Doczi, S. Komoly and A. Buki, *Acta Neurochir. (Wien)*, 2010, **152**(1), 1.
23. B. K. Kwon, A. M. Stammers, L. M. Belanger, A. Bernardo, D. Chan, C. M. Bishop, G. P. Slobogean, H. Zhang, H. Umedaly, M. Giffin, J. Street, M. C. Boyd, S. J. Paquette, C. G. Fisher and M. F. Dvorak, *J. Neurotrauma*, 2010, **27**(4), 669.
24. B. K. Kwon, S. Casha, R. J. Hurlbert and V. W. Yong, *Clin. Chem. Lab. Med.*, 2011, **49**(3), 425.
25. L. Pellerin and P. J. Magistretti, *Proc. Natl. Acad. Sci. USA*, 1994, **91**, 10625.
26. G. Brasnjo and T. S. Otis, *Proc. Natl. Acad. Sci. USA*, 2004, **101**, 6273.

27. A. Panov, P. Schonfeld, S. Dikalov, R. Hemendinger, H. Bonkovsky and B. R. Brooks, *J. Biol. Chem.*, 2009, **284**, 14448.
28. K. R. Dave, W. G. Bradley and M. A. Perez-Pinzon, *Exp. Neurol.*, 2003, **182**, 412.
29. P. G. Sullivan, A. G. Rabchevsky, J. Y. N. Keller, M. Lovell, A. Sodhi, R. P. Hart and S. W. Scheffer, *J. Comparative Neurol.*, 2004, **474**, 524.
30. F. Forner, L. J. Foster, S. Campanaro, G. Valle and M. Mann, *Mol. Cell. Proteomics*, 2006, **5**, 608.
31. C. Krieger and M. R. Duchen, *Eur. J. Pharmacol.*, 2002, **447**, 177.
32. I. J. Reynolds, *Ann. N.Y. Acad. Sci.*, 1999, **893**, 33.
33. S. Chalmers and D. G. Nicholls, *J. Biol. Chem.*, 2003, **278**, 19062.
34. A. Panov, L. Andreeva and J. T. Greenamyre, *Arch. Biochem. Biophys.*, 2004, **424**, 44.
35. A. Panov, S. Dikalov, N. Shalbuyeva, R. Hemendinger, J. T. Greenamyre and J. Rosenfeld, *Am. J. Physiol. Cell. Physiol.*, 2007, **292**, C708.
36. A. P. Somlyo and A. V. Somlyo, *J. Cardiovasc. Pharmacol.*, 1986, **8**, S42.
37. D. G. Nicholls, *Curr. Mol. Med.*, 2004, **4**, 149.
38. http://www.sigmaaldrich.com/etc/medialib/docs/Sigma/Formulation/d5546for.Par.0001.File.tmp/d5546for.pdf.
39. J. W. Rowland, G. W. Hawryluk, B. Kwon and M. G. Fehlings, *Neurosurg. Focus*, 2008, **25**(5), E2.
40. A. P. Halestrap, E. Doran, J. P. Gillespie and A. O'Toole, *Biochem. Soc. Trans.*, 2000, **28**, 170.
41. D. F. Stowe and A. K. Camara, *Antioxidants & Redox Signaling*, 2009, **11**(6), 1381.
42. A. A. Starkov, *Ann. New York Acad. Sci.*, 2008, **1147**, 37.
43. F. Zoccarato, L. Cavallini, S. Bortolamiand and A. Alexandre, *Biochem. J.*, 2007, **406**(1), 125.
44. A. Panov, N. Steuerwald, V. Vavilin, S. Dambinova and H. L. Bonkovsky, in *Amyoptrophic Lateral Sclerosis*, ed. M. H. Maurer, InTech Publishing, Croatia, 2012, pp. 225–248. Free online edition is available at www.intechopen.com
45. R. Balazs, *J. Neurochem.*, 1965, **12**, 63.
46. R. Balazs, *Biochem. J.*, 1965, **95**, 497.
47. N. J. Tillakaratne, L. Medina-Kauweand and K. M. Gibson, *Comp. Biochem. Physiol. A Physiol.*, 1995, **112**(2), 247.
48. M. T. Wong-Riley, *Trends Neurosci.*, 1989, **12**(3), 94.
49. L. Hertz, *J. Cereb. Blood Flow Metab.*, 2004, **24**(11), 1241.
50. T. V. Votyakova and I. J. Reynolds, *J. Neurochem.*, 2001, **79**, 266.
51. E. Park, A. A. Velumian and M. G. Fehlings, *J. Neurotrauma*, 2004, **21**, 754.
52. R. L. Blaylock and J. Maroon, *Surg. Neurol. Intern.*, 2011, **2**, 107.
53. W. Deng, P. A. Rosenberg, J. J. Volpe and F. E. Jensen, *Proc. Natl. Acad. Sci. USA*, 2003, **100**(11), 6801.
54. P. K. Stys and S. A. Lipton, *Trends Pharmacol. Sci.*, 2007, **28**(11), 561.
55. R. F. Regan, *Neurosci. Lett.*, 1996, **213**(1), 9.

56. J. R. Wrathall, D. Choiniere and Y. D. Teng, *J. Neurosci.*, 1994, **14**(11, Pt 1), 6598.
57. I. Micu, Q. Jiang, E. Coderre, A. Ridsdale, L. Zhang, J. Woulfe, X. Yin, B. D. Trapp, J. E. McRory, R. Rehak, G. W. Zamponi, W. Wang and P. K. Stys, *Nature*, 2006, **439**(7079), 988.
58. P. K. Stys, *Curr. Mol. Med.*, 2004, **4**(2), 113.
59. M. O. Totoiu and H. S. Keirstead, *J. Comp. Neurol.*, 2005, **486**(4), 373.
60. S. Baltan, *Neuroscientist*, 2009, **15**(2), 126.
61. E. Esposito, I. Paterniti, E. Mazzon, T. Genovese, M. Galuppo, R. Meli, P. Bramanti and S. Cuzzocrea, *BMC Neurosci.*, 2011, **12**, 31.
62. R. D. Azbill, X. Mu, A. J. Bruce-Keller, M. P. Mattson and J. E. Springer, *Brain Res.*, 1997, **765**(2), 283.
63. G. Fiskum, *J. Neurotrauma*, 2000, **17**(10), 843.
64. A. Lewen, P. Matz and P. H. Chan, *J. Neurotrauma*, 2000, **17**(10), 871.
65. D. Liu, J. Liu, D. Sun and J. Wen, *J. Neurotrauma*, 2004, **21**(6), 805.
66. E. Park, A. A. Velumian and M. G. Fehlings, *J. Neurotrauma*, 2004, **21**, 754.
67. M. C. Tsai, C. P. Wei, D. Y. Lee, Y. T. Tseng, M. D. Tsai, Y. L. Shih, Y. H. Lee, S. F. Chang and S. J. Leu, *Surg. Neurol.*, 2008, **70**(S1), 19.
68. H. Ouyang, W. Sun, Y. Fu, J. Li, J. X. Cheng, E. Nauman and R. Shi, *J. Neurotrauma*, 2010, **27**(6), 1109.
69. G. Thibault-Halman, S. Casha, S. Singer and S. Christie, *J. Neurotrauma*, 2011, **28**(8), 1497.
70. V. M. Runge, J. W. Wells, S. A. Baldwin, S. W. Scheff and D. A. Blades, *Invest. Radiol.*, 1997, **32**(2), 105.
71. J. M. Lubieniecka, F. Streijger, J. H. Lee, N. Stoynov, J. Liu, R. Mottus, T. Pfeifer, B. K. Kwon, J. R. Coorssen, L. J. Foster, T. A. Grigliatti and W. Tetzlaff, *PLoS One*, 2011, **6**(4), e19247.
72. A. Raabe, C. Grolms and V. Seifert, *Br. J. Neurosurg.*, 1999, **13**(1), 56.
73. R. D. Rothoerl, C. Woertgen, M. Holzschuh, C. Metz and A. Brawanski, *J. Trauma*, 1998, **45**(4), 765.
74. B. Romner, T. Ingebrigtsen, P. Kongstad and S. E. Borgesen, *J. Neurotrauma*, 2000, **17**(8), 641.
75. V. Pancholi, *Cell. Mol. Life Sci.*, 2001, **58**(7), 902.
76. L. Pellerin, A. K. Bouzier-Sore, A. Aubert, S. Serres, M. Merle, R. Costalat and P. J. Magistretti, *Glia*, 2007, **55**(12), 1251.
77. R. George and J. W. Griffin, *Exp. Neurol.*, 1994, **129**(2), 225.
78. R. Donato, *Cell Calcium*, 1986, **7**(3), 123.
79. R. Donato, B. Prestagiovanni and G. Zelano, *J. Neurochem.*, 1986, **46**(5), 1333.
80. M. Mata, D. Alessi and D. J. Fink, *J. Neurocytol.*, 1990, **19**(3), 432.
81. G. R. Campbell and D. J. Mahad, *Autoimmune Dis.*, 2011, Article ID 262847.
82. A. P. Halestrap, *Nature*, 2005, **434**, 578.
83. M. Abeles, in *Corticonics: Neural Circuits of the Cerebral Cortex*, Cambridge University Press, 1991.

84. D. Attwell and S. B. Laughlin, *J. Cereb. Blood. Flow. Metab.*, 2001, **21**, 1133.
85. P. G. Sullivan, S. Krishnamurthy, S. P. Patel, J. D. Pandya and A. G. Rabchevsky, *J. Neurotrauma*, 2007, **24**(6), 991.
86. R. D. Azbill, X. Mu, A. J. Bruce-Keller, M. P. Mattson and J. E. Springer, *Brain Res.*, 1997, **765**(2), 283.
87. T. Ito, N. Allen and D. Yashon, *J. Neurosurg.*, 1978, **48**(3), 434.
88. S. D. Christie, B. Comeau, T. Myers, D. Sadi, M. Purdy and I. Mendez, *Neurosurg. Focus*, 2008, **25**(5), E5.
89. D. Liu, J. Liu, D. Sun and J. Wen, *J. Neurotrauma*, 2004, **21**(6), 805.
90. A. Lewen, P. Matz and P. H. Chan, *J. Neurotrauma*, 2000, **17**(10), 871.
91. G. J. Zipfel, D. J. Babcock, J. M. Lee and D. W. Choi, *J. Neurotrauma*, 2000, **17**(10), 857.

CHAPTER 7

Biomarkers of Neuroglial Injury in Rat Models of Combat TBI: Primary Blast Over-Pressure Compared to "Composite" Blast

STANISLAV I. SVETLOV,[*a,c] VICTOR PRIMA,[*a]
OLENA GLUSHAKOVA,[a] ARTEM SVETLOV,[a]
DANIEL R. KIRK,[b] HECTOR GUTIERREZ,[b]
KEVIN K. W. WANG[d] AND RONALD L. HAYES[a]

[a] Banyan Laboratories, Inc., 12085 Research Drive, Alachua, FL 32615,
USA; [b] Department of Mechanical and Aerospace Engineering, Florida
Institute of Technology, Melbourne, FL, USA; [c] Department of Medicine,
University of Florida, Gainesville, FL, USA; [d] Department of Psychiatry,
University of Florida, Gainesville, FL, USA
*Email: ssvetlov@banyanbio.com and vprima@banyanbio.com

7.1 Introduction

The nature of 21st century warfare has led to a significant increase in human exposure to blast over-pressure (OP) impulses, which result in a complex of neurosomatic disorders, including traumatic brain injury (TBI). Blast-related casualties have outnumbered conventional injuries during the past several years in Iraq and Afghanistan, while blast itself is being termed "the fourth weapon of mass destruction."[1] Moreover, for every blast-related fatality, many more soldiers suffer multiple, low level non-lethal blast exposures. This often leads to mild traumatic brain injury (mTBI), which is rarely recognized in a timely

RSC Drug Discovery Series No. 24
Biomarkers for Traumatic Brain Injury
Edited by Svetlana A. Dambinova, Ronald L. Hayes and Kevin K. W. Wang
© Royal Society of Chemistry 2012
Published by the Royal Society of Chemistry, www.rsc.org

manner and has become a signature injury of the Iraq and Afghanistan conflicts.[2–4] Symptoms of mild or moderate blast brain injury often do not manifest themselves until sometime after the injury has occurred[5–8] and go undiagnosed and untreated because emergency medical attention is directed towards more visible injuries, such as penetrating flesh wounds.[9–11] However, even mild and moderate brain injuries can produce significant deficits and, particularly when repeated, can lead to sustained neurosomatic damage and neurodegeneration.[6] Thus, identifying pathogenic mechanisms and biochemical markers of blast brain injury in relevant experimental models is vital to the development of diagnostics for mTBI through severe TBI.

However, because of the design inconsistency of blast/shock generators used in different studies, incomplete understanding of blast wave biophysics associated with real explosives versus those produced by air or gas-driven shock tubes, and details of wave interaction with model animals, disparities between laboratory models and data on brain injury mechanisms and putative biomarkers have been difficult to analyze and compare[12–15] (see Bass *et al.*[16] for a review). Moreover, pathogenic pathways and molecular signatures of neural responses and injurious effects of blast exposures remain elusive. Recently, we developed and employed a model of "composite" blast exposure with controlled parameters of blast wave impact and brain injury in rats.[17] Our studies demonstrated the importance of positional orientation of the head and whole body of rats toward a blast wave generated from an external shock tube.[18] Data from several laboratories, including our studies,[17,18] suggest that the mechanisms underlying blast-induced injuries, particularly mild/moderate, appear to be distinct from those imposed by mechanical impact or acceleration, and may involve the prominent systemic response (see Cernak[19] for a review).

In this study, we compared the effects of moderate primary over-pressure (OP) body/head exposure with brain injury produced by a severe blast accompanied by strong head acceleration. The high speed imaging using Schlieren optics demonstrated blast wave interaction with the animal's head/body and revealed a negligible degree of acceleration at a position "off-axis" with the shock tube (primary blast wave exposure) compared to the "on-axis" experimental setup, which was accompanied by strong head/cervical acceleration generated by peak OP + venting gas (composite blast, or primary blast wave plus gas jetting phenomena). The specific dynamics of systemic, vascular, inflammatory, and neuroglial injury signatures, including NSE/UCH-L1, GFAP, and CNPase biomarkers in serum, were established and characterized.

7.2 Materials and Methods

7.2.1 Hardware Design and Setup

The compressed air-driven shock tube capable of generating a wide range of controlled blast waves has been previously described. The tube consists of two sections: high-pressure (driver) and low-pressure (driven) separated by a diaphragm. Peak over-pressure, composition, and duration of the high pressure shockwaves that are generated are determined by the shock tube configuration, including thickness, type of diaphragm material, driver/driven ratio, and the initial driver pressure at the moment of diaphragm rupture. In the series of experiments designed to explore the effects of different components of the blast/ shock waves on the targeted animal brain, we employed different spatial setups, as described below. The blast pressure data was acquired using PCB piezo-electric blast pressure transducers and LabView 8.2 software. A National Instruments 1.25 M samples/sec data acquisition card was used to acquire data from multiple channels. The rat head images during the blast event were captured at 40 000 frames/second using a high speed video camera (Phantom V310, Vision Research, Wayne, NJ) and mirror-based Schlieren optics.

7.2.2   Animal Exposure to a Controlled Blast Wave

Modeling of the primary blast and the "composite" over-pressure load was achieved by variable positioning of the target versus the blast generator. All rats were anesthetized with isoflurane inhalations described previously in detail. After reaching a deep plane of anesthesia, they were placed into a holder exposing either only their head (body-armored setup) or whole body at the distance 5 cm below the exit nozzle of the shock tube. Rats were positioned either directly on the shock tube axis or at a 45° angle to it to expose them correspondingly to the "composite" blast including the compressed air jet or only to the primary blast wave (see Figures 7.1D and 7.1E). Animals were then subjected to a single blast with a mean peak over-pressure at the target of 230–380 kPa (see Figures 7.1A and 7.1B). The exact static and dynamic over-pressure values depending on the angle and distance of rat head from the nozzle of shock tube were established during the prior calibration tests (see Figure 7.1C). The control group of animals underwent the same treatment (anesthesia, handling, recovery) except they were not exposed to the blast.

7.2.3   Blood and Tissue Collection

At the required time points following blast exposure, animals were euthanized according to guidelines approved by the IACUC of the University of Florida. Blood was withdrawn directly from the heart under isoflurane anesthesia and brain tissue samples were collected, snap-frozen in liquid nitrogen, and stored at −70 °C until further analysis.

At 1 and 7 days after TBI (primary, head-only blast) animals were euthanized with lethal dose of pentobarbital, transcardially perfused with 4% paraformaldehyde, and whole brains were removed, processed, and embedded in paraffin. IHC analysis was performed on paraffin-embedded 6 μm brain sections. Slides were de-paraffinized, incubated for 10 min at 95 °C in Trilogy solution (Cell Marque, Rocklin, CA) for antigen retrieval, blocked for endogenous peroxides, and incubated with primary antibodies for GFAP (Cell Signaling Technology, Danvers, MA) or CNPase (Abcam, Cambridge, MA) overnight at 4 °C followed by treatments with secondary antibodies.

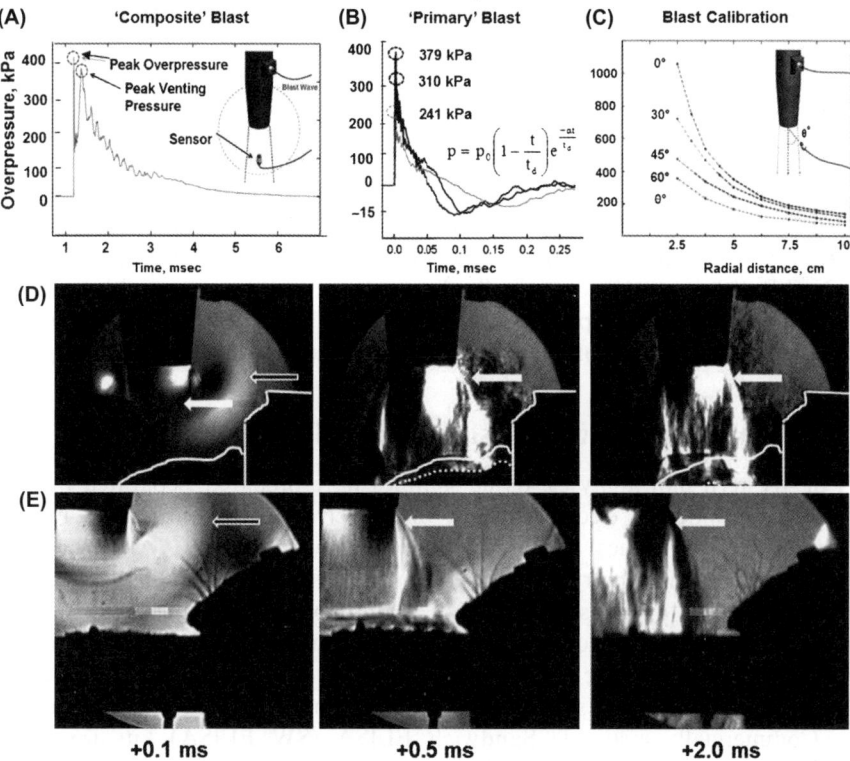

Figure 7.1   Schematic presentation of blast exposure modeling and blast visualization in rats using Schlieren optics. A: overpressure recording on shock tube axis at 5 cm from the nozzle. B: overpressure recording with external "pencil" PCB at 5 cm and 45° from shock tube nozzle at three different diaphragm configurations. Inset formula in B: an empirical expression for the pressure decay with time at a fixed distance is characterized by a decay parameter α (G. F. Kinney, *Explosive Shocks in Air*, New York, Springer-Verlag, 1985). C: calibration of pressure on rat head depending on the angle and distance from the nozzle of shock tube. D and E: different shock tube set-ups to model "primary" and "composite" blast. D: "composite blast". E: "primary blast." Black arrows indicate formation, traveling, and interaction of blast wave with rat head (accomplished within ∼ 0.1 ms). White arrows show gas venting jet hitting rat head after blast wave passed through (persists for milliseconds). The solid contour line in panel A outlines the shape of animal head at time point 0; the dotted line shows the current shape. Please see Materials and Methods for details.

The staining was visualized with 3,3'-diaminobenzidine (DAB) (Dako, Carpinteria, CA) for brown color development. Sections were counterstained with Hematoxylin (Dako). Negative controls were performed by treatment with species-matched secondary antibodies only (not shown). The slides were scanned and examined using Aperio ScanScope GL system with either 5× or 20× objective and ScanScope software.

7.2.4   Silver Staining Assessment of Neurodegeneration in Rat Brain

Neuroinjury and neurodegeneration were examined in the perfused and fixed brains using silver staining histochemical procedures according to Neuroscience Associates (Knoxville, TN), utilizing the de Olmos Amino Cupric Silver Stain previously described in detail.[17] In addition, the silver staining kit from FD NeuroTechnologies (Ellicott City, MD) was used where indicated. Rats were subjected to (i) "composite" head-directed severe blast exposure (358 kPa/10 ms total) on-axis (body protected); (ii) primary blast off-axis exposure to peak over-pressure only (233 kPa/113 μs total); and (iii) controlled cortical impact (CCI) of 2.0 mm depth performed as described previously.[20]

7.2.5   Western Blot Analysis of Brain Tissues

For Western blot analyses, tissue samples were prepared, separated by SDS-polyacrylamide gel electrophoresis, and electro-blotted onto polyvinylidene difluoride membranes as described previously in detail.[17] After overnight incubation with primary antibodies for CNPase (Cell Signaling Technology, Danvers, MA), proteins were incubated with conjugated secondary antibodies and detected by either colorimetric or chemiluminescent (ECL) detection systems. Semi-quantitative assessment of protein levels by Western blot densitometry was conducted using the NIH ImageJ image processing program.

Commercially available Sandwich ELISA (SW ELISA) kits for GFAP (BioVendor, Candler, NC), and NSE (Life Sci. Advanced Tech., St. Petersburg, FL), were used according to the manufacturer's instructions. Ubiquitin C-terminal hydrolase L1 (UCH-L1) in CSF and plasma was quantitatively detected using proprietary SW ELISA (Banyan Biomarkers, Inc.) and recombinant UCH-L1 as standard.

7.2.6   Statistics

Statistical analyses were performed using GraphPad Prism 5 software. Values are means $\pm$ SEM. Data were evaluated by 2-tailed unpaired t-test with or without Welch corrections where indicated.

7.3   Results and Discussion

7.3.1   Rat Models of Blast Exposure Using External Shock Tube: Primary Blast Load versus "Composite" Blast Exposure

Our shock tube was designed and built to model a freely expanding blast wave as generated by a typical explosion. Both static and dynamic (total) pressures were measured as functions of angle and radial distance from shock tube exit using piezoelectric blast pressure transducers positioned at the target (see Figure 7.1C). The pressure transducers registered three distinct events: (i) peak OP, (ii) gas venting jet, on axis only, and (iii) negative pressure phase, off axis only (see Figures 7.1A and 7.1B). The exhaust of venting gas apparently distorted the propagation of the blast wave and no negative phase was registered when dynamic pressure was measured on axis of the shock tube (see Figure 7.1A), while a distinct and substantial negative phase (15–20 kPa) was detected off axis (see Figure 7.1B). Peak OP, positive phase duration, and impulse appear to be the key parameters that correlate to injury and likelihood of fatality in animals and humans for various orientations of the specimen relative to the blast wave. Shock tubes produce a "venting gas jet" immediately after the blast wave forms, substantially contaminating the blast wave in the direction of the shock tube axis. In a composite blast setup, the venting gas jet lasts the longest (up to ~3–5 ms), albeit lower in magnitude than peak OP, represents the bulk of blast impulse, and possibly produces the most devastating impact. Schlieren optics (see Figure 7.1D) demonstrated a strong downward head acceleration following the passage of peak OP which lasts 50–100 μs. However, cranial deformation was more severe during the gas venting phase, lasting up to 5 ms. This effect was eliminated by placing rats off-axis from the venting jet in a way that the main effect acting on the specimen is the peak OP event. The high speed recording coupled with Schlieren optical system visualized interaction of the blast wave with the animal head/body and revealed a negligible degree of acceleration at rat positioning "off-axis" toward shock tube (primary blast) (see Figure 7.1E). The pressure on the surface of rats was calibrated depending on the distance and angle from the nozzle of shock tube (see Figure 7.1C).

Shock tubes have been used as the fundamental research tool for the last several decades.[13–15,21] There is still concern whether a blast wave generated by shock tubes using compressed gas accurately reflects a real explosive blast. In our study, dynamic pressure measured by a PCB "pencil" sensor indicated that shock tubes produce a "venting gas jet" immediately after blast wave formation (see the shoulder at Figure 7.1A), substantially contaminating the blast wave in the direction of shock tube axis (see Figure 7.1A). In addition, the exhaust "venting gas" apparently masked the negative phase of the shock wave, which was present when the dynamic pressure was recorded at an angle to the shock tube nozzle (see Figure 7.1B). Schlieren optics techniques clearly defined the areas of pressure, either peak OP or venting gas jet (see Figures 7.1D and 7.1E). This pattern is characteristic of "external" shock tube models, where the target/animal is placed outside rather than within the tube. Placing animals within the tube also can produce confounding effects when the animal is very large relative to the tube diameter or when the animal is suspended and or shielded inappropriately. The shape of the blast wave and the development of constructive or destructive secondary waves as the primary wave exits the tube can be affected by the size and shape of the exit as well. This can be visualized with Schlieren optics. By placing rats off-axis from the shock tube nozzle, we eliminated the venting gas in a way that the main effect acting on the rat is the peak OP event and the negative phase of the blast wave. Thus, we examined the pathological impact of two different types of blast with precisely controlled magnitude, duration, and impulse at the surface of the rat, different orientations of the head to the blast wave, and open or armored body: (i) primary blast/peak OP only with rats located off-axis with the shock tube and (ii) composite blast with rats located on axis, accompanied by linear and, to a lesser extent, rotational head hyperacceleration (see Figure 7.1D). It should be noted that any blast produced in the laboratory models only a particular component of a complex blast that might be experienced on the battlefield. The detonation of real explosives in the field does not produce the "venting gas", but can result in significant bulk flow of air and debris. This makes the separation of the effects of primary and particularly tertiary blast (the target being displaced by the blast) difficult to separate in most existing testing regimens. Although the blast generated in our on-axis model is a single blast event, the type of blast load observed resembles the complex effect produced by multiple blasts, such as in a confined space where the blast waves reverberate and overlap; hence, the effect of displaced air mass flow on the resultant wave structure and magnitude can be important.

7.3.2 Neural Injury and Gliosis in Rat Brain after Different Blast Exposures Assessed by Silver Staining and Immunohistochemistry (IHC)

As shown in Figure 7.2, composite blast (on-axis) produces silver accumulation at the 7th day post blast (see Figures 7.2A and 7.2D), particularly in the hippocampus (indicated by arrows). Controlled cortical impact (CCI) also results in positive staining in ipsilateral cortex and hippocampus (see Figures 7.2C and 7.2F). In contrast, there was a rare occurrence of silver accumulation observed in the cortex or hippocampus after exposure to primary blast (see Figures 7.2B and 7.2E) (indicated by arrows).

There was a substantial difference in the effects of composite versus primary blast on neurodegenerative processes in the cortex; particularly in the hippocampus at 7 days post blast (see Figure 7.2). Silver accumulation in the cortex after composite blast was modest, with a very rare finding of "classical type" neurodegeneration (see Figure 7.2A, inset). On the other hand, the hippocampus significantly accumulated silver in fiber-like structures after composite blast (see Figure 7.2D), while very occasional silver staining was observed in both cortex and hippocampus after primary blast (see Figures 7.2B and 7.2E). As expected and in accordance with data reported previously, controlled cortical impact evoked a distinct cellular neurodegeneration in both cortical and hippocampal tissue (see Figures 7.2C and 7.2F). The most common types of closed head impact TBI are diffuse axonal injury, contusion, and subdural hemorrhage as an overall result of rotational acceleration.[22] Diffuse axonal injuries are very common following closed head injuries. They result when shearing, stretching, and/or angular forces pull on axons and small

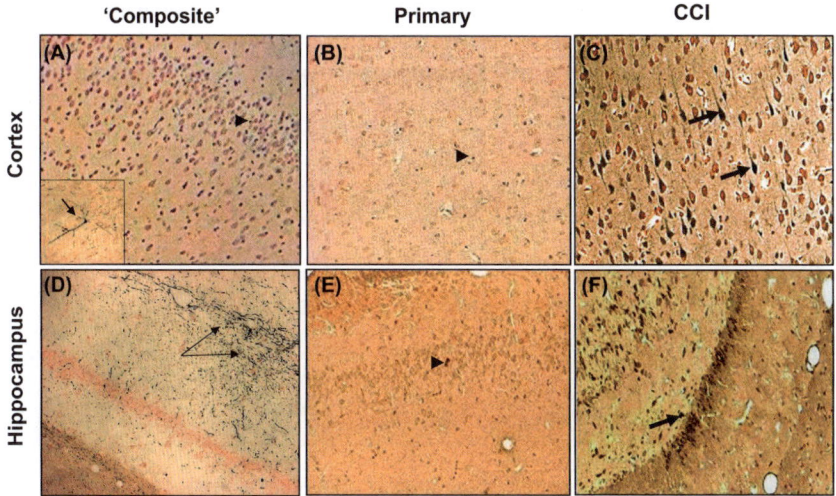

Figure 7.2   Silver staining of coronal brain sections following primary or "composite" blast exposure. Corresponding tissue staining 7 days after "composite blast," primary blast, and CCI is shown in panels A, B, and C for cortex, and in panels D, E, and F for hippocampus. Arrowheads indicate occasional silver accumulation in the cells of non-neuronal origin. Arrows indicate diffuse silver accumulation in neurons. A inset: a very rare accumulation of silver in a cortical neuron. Please see Materials and Methods for details.

vessels. Impaired axonal transport leads to focal axonal swelling and, after several hours, may result in axonal disconnection.[23] The typical locations are the corticomedullary (gray matter/white matter) junction, internal capsule, deep gray matter, upper brainstem, and corpus callosum. Multifocal axonal degeneration, as evidenced by amino cupric silver staining, is characteristic also for shock wave insult, as was shown in a study with head-only exposed rats inside a shock tube.[24] Our recent[17] and current studies clearly demonstrate the presence of neural degeneration in deeper structures of the brain; specifically, the hippocampus after composite blast producing linear and rotational head acceleration, which is lacking or negligible following primary blast.

Time-dependent expression of GFAP and CNPase characteristic for astrocytes and oligodendrocytes, respectively, was studied by IHC after moderate composite on-axis blast (358 kPa/~10 ms) with strong head acceleration or moderate primary off-axis blast (234 kPa/113.8 µs positive phase) with minor head acceleration (see Figure 7.3). These data suggest that both primary and "composite" blasts strongly induce astrogliosis (GFAP; see Figure 7.3, upper panel) and oligodendrocytosis (CNPase; see Figure 7.3, lower panel) in rat hippocampus evident as early as day 1 and lasting up to 7 days post blast. Markers of activated astrocytes GFAP and oligodendrocytes CNPase were strongly up-regulated in CA1 and DG regions of the hippocampus, respectively, at day 1 and sustained up to 7 days post blast. These findings are in strict

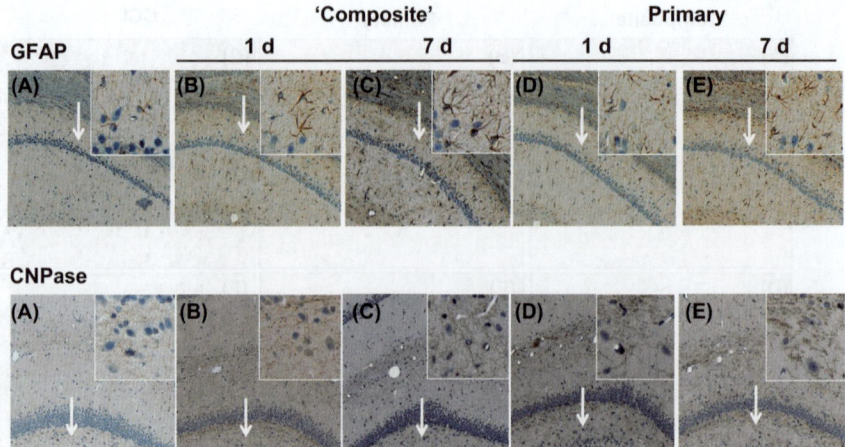

Figure 7.3   Immunohistochemical analysis of astrocyte and oligodendrocyte markers in hippocampus after blast. Time-dependent GFAP and CNPase expression was studied by IHC on paraffin-embedded 6 μm brain sections after blast exposure at different set-ups. A: naive; B: "composite" blast, 1 day; C: "composite", 7 days; D: "primary", day 1 and E: "primary", 7 days. Magnifications 5× and 20× (insets) are shown. Arrows indicate inset locations for CA1 region (GFAP) and DG region (CNPase) in hippocampus. Please see Materials and Methods for details.

accordance with many previous reports, including from our group, supporting the notion that gliosis represents a common and rapid response to brain insult regardless of the nature: mechanical or blast-induced exposure.[17,25,26]

7.3.3   Serum Levels of Biomarkers of Neuroglial Injury Following Blast Exposure

To assess if markers of neuronal injury are released into circulation, we assayed serum levels of neuron-specific enolase (NSE) and UCH-L1 after different blast exposures (see Figures 7.4A and 7.4B). As shown in Figure 7.4A, remarkable accumulation of NSE in serum occurred within 6 hours following exposure to either "composite" or primary blast, and persisted up to 7 days post blast. Average serum UCH-L1 level was also elevated during 1–7 days after "primary" blast (see Figure 7.4B), although its difference from controls was statistically significant only at day 1 post blast.

Glia cell-specific up-regulation of GFAP and CNPase in the brain after either "composite" or primary blast was accompanied by a significant serum accumulation of GFAP and CNPase biomarkers measured by SW ELISA for GFAP (see Figure 7.4C) and semi-quantitative Western blot densitometry for serum CNPase (see Figure 7.4D). These biomarkers persisted in blood up to 7 days post blast at both blast setups employed.

In these experiments, we used NSE SW ELISA Kit from Life Sciences Advanced Technologies designed specifically to detect rat NSE. However, several reports indicate that NSE may not be highly specific for the CNS and is present in platelets and red blood cells (see Svetlov *et al.*[27] for a review). In previous studies, we reported a slight UCH-L1 increase after "composite" blast, followed by a rapid decline.[17] The UCH-L1 SW ELISA used in early experiments had low specificity and sensitivity for rat samples; thus, many serum substances interfered and masked the UCH-L1 content. In this study, an improved version of the UCH-L1 assay was employed, which was still not particularly specific for rats (data not shown). Increases in serum UCH-L1 were statistically significant only at day 1 after a single primary blast exposure, although an elevation trend could be detected (see Figure 7.4B). In contrast, a rat-specific GFAP SW ELISA has been generated and employed in these studies. Serum GFAP increase was prominent within 6 hours after composite and

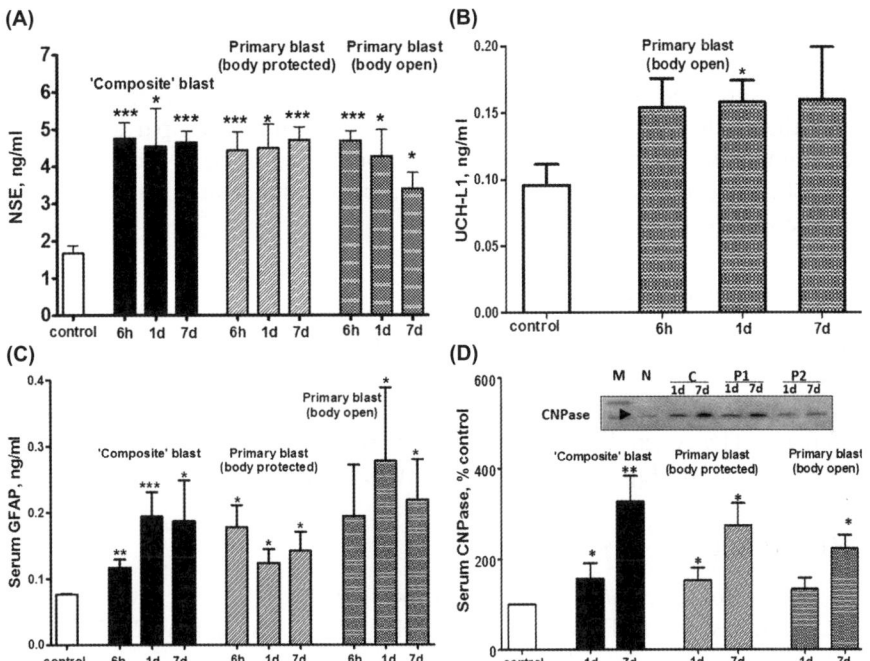

Figure 7.4 Blast-induced accumulation of NSE, UCHL-1, GFAP, and CNPase in rat serum. Blood was collected from overpressure-exposed rats at different shock tube setups and assayed by NSE (A), UCHL-1 (B), and GFAP (C) by SW ELISA Kits. (D): semi-quantitative serum CNPase detection by western blot densitometry. Inset: representative western blot. N = naïve; C = "composite"; P1 = primary/head; P2 = primary/body). t-test with Welch correction was done. Data shown are mean + SEM of at least three independent experiments. *P < 0.05; **P < 0.01; ***P < 0.005 versus naïve. Please see Materials and Methods for details. Unpaired *t*-test was used to analyze statistical significance of values.

primary blast with body protected (see Figure 7.4C), and elevated levels persisted up to 7 days post blast, consistent with up-regulation in hippocampus. The CNPase content assessed by semi-quantitative Western blot was raised at day 1 after blast exposure (except primary blast with open body) and further substantially increased at 7 day post blast (see Figure 7.4D). It remains to be examined whether CNPase up-regulation reflects a long-term disorder of myelination following blast exposure and whether CNPase can be a biomarker of chronic injury.

7.4   Conclusions

In conclusion, the specific dynamics of systemic, vascular inflammatory, neuroendocrine, growth factor, and neuroglial biomarkers in serum were established and characterized. For major pathway signatures and biomarkers, detected levels were increased at all the setups studied. However, the most significant and persistent changes in neuroglial injury markers were found after composite blast, while primary blast instigated the most prominent systemic/vascular, neuroendocrine, and growth factor responses, particularly when the rat was subjected to frontal, head-directed, open body exposure. We suggest that the mechanisms underlying primary blast brain injuries, particularly mild and moderate, are different from blast accompanied by head acceleration and may be triggered by systemic, cerebrovascular, and neuroglial responses as overlapping events.

Acknowledgements

The authors wish to thank Mr. Danny Johnson for his excellent technical assistance. This work was supported by grants W81XWH-8-1-0376, W81XWH-10-1-0876 and W81XWH-07-01-0701 from the U.S. Department of Defense. The research was conducted in the absence of any commercial or financial relationships that could be construed as a potential conflict of interest.

References

1. C. T. Born, *Scand. J. Surg.*, 2005, **94**(4), 279.
2. E. Jones, N. T. Fear and S. Wessely, *Am. J. Psychiatry*, 2007, **164**(11), 1641.
3. H. Terrio, L. A. Brenner, B. J. Ivins, J. M. Cho, K. Helmick, K. Schwab, K. Scally, R. Bretthauer and D. Warden, *J. Head Trauma Rehabil.*, 2009, **24**(1), 14.
4. D. Warden, *J. Head Trauma Rehabil.*, 2006, **21**(5), 398.
5. I. Cernak, A. C. Merkle, V. E. Koliatsos, J. M. Bilik, Q. T. Luong, T. M. Mahota, L. Xu, N. Slack, D. Windle and F. A. Ahmed, *Neurobiol. Dis.*, 2011, **41**(2), 538.
6. I. Cernak and L. J. Noble-Haeusslein, *J. Cereb. Blood Flow Metab.*, 2010, **30**(2), 255.

7. I. Cernak, J. Savic, D. Ignjatovic and M. Jevtic, *J. Trauma*, 1999, **47**(1), 96, discussion 103.
8. S. Yilmaz and M. Pekdemir, *Am. J. Emerg. Med.*, 2007, **25**(1), 97.
9. H. G. Belanger, S. G. Scott, J. Scholten, G. Curtiss and R. D. Vanderploeg, *J. Rehabil. Res. Dev.*, 2005, **42**(4), 403.
10. T. J. Nelson, D. B. Wall, E. T. Stedje-Larsen, R. T. Clark, L. W. Chambers and H. R. Bohman, *J. Am. Coll. Surg.*, 2006, **202**(3), 418.
11. S. J. Wolf, V. S. Bebarta, C. J. Bonnett, P. T. Pons and S. V. Cantrill, *Lancet*, 2009, **374**(9687), 405.
12. M. Chavko, S. Adeeb, S. T. Ahlers and R. M. McCarron, *Shock*, 2009, **32**(3), 325.
13. N. M. Elsayed, *Toxicology*, 1997, **121**(1), 1.
14. R. J. Guy, E. Kirkman, P. E. Watkins and G. J. Cooper, *J. Trauma*, 1998, **45**(6), 983.
15. J. H. Jaffin, L. McKinney, R. C. Kinney, J. A. Cunningham, D. M. Moritz, J. M. Kraimer, G. M. Graeber, J. B. Moe, J. M. Salander and J. W. Harmon, *J. Trauma*, 1987, **27**(4), 349.
16. C. R. Bass, M. B. Panzer, K. A. Rafaels, G. Wood, J. Shridharani and B. Capehart, *Ann. Biomed. Eng.*, 2011 Oct 20 (Epub ahead of print).
17. S. I. Svetlov, V. Prima, D. R. Kirk, H. Gutierrez, K. C. Curley, R. L. Hayes and K. K. Wang, *J. Trauma*, 2010, **69**(4), 795.
18. S. I. Svetlov, V. Prima, R. L. Hayes, K. K. K. Wang, D. R. Kirk, H. Gutierrez and K. C. Curley, *Proceedings of NATO Conference*, RTO-MP-HFM-207, 2011; paper 37. http://www.rto.nato.int
19. I. Cernak, *Front. Neurol.*, 2010, **1**, 151.
20. M. C. Liu, L. Akinyi, D. Scharf, J. Mo, S. F. Larner, U. Muller, M. W. Oli, W. Zheng, F. Kobeissy, L. Papa, X. C. Lu, J. R. Dave, F. C. Tortella, R. L. Hayes and K. K. Wang, *Eur. J. Neurosci.*, 2010, **31**(4), 722.
21. R. J. Guy, M. A. Glover and N. P. Cripps, *J. R. Nav. Med. Serv.*, 1998, **84**(2), 79.
22. M. Vander Vorst, K. Ono, P. Chan and J. Stuhmiller, *J. Trauma*, 2007, **62**(1), 199.
23. R. A. Hurley, J. C. McGowan, K. Arfanakis and K. H. Taber, *J. Neuropsychiatry Clin. Neurosci.*, 2004, **16**(1), 1.
24. R. H. Garman, L. W. Jenkins, R. C. Switzer, R. A. Bauman, L. C. Tong, P. V. Swauger, S. A. Parks, D. V. Ritzel, C. E. Dixon, R. S. Clark, H. Bayir, V. Kagan, E. K. Jackson and P. M. Kochanek, *J. Neurotrauma*, 2011, **28**(6), 947.
25. S. K. Kwon, E. Kovesdi, A. B. Gyorgy, D. Wingo, A. Kamnaksh, J. Walker, J. B. Long and D. V. Agoston, *Front. Neurol.*, 2011, **2**, 12.
26. C. Urrea, D. A. Castellanos, J. Sagen, P. Tsoulfas, H. M. Bramlett and W. D. Dietrich, *Restor. Neurol. Neurosci.*, 2007, **25**(1), 65.
27. S. I. Svetlov, S. F. Larner, D. R. Kirk, J. Atkinson, R. L. Hayes and K. K. Wang, *J. Neurotrauma*, 2009, **26**(6), 913.

CHAPTER 8

Biomarkers for Subtle Brain Dysfunction

SVETLANA A. DAMBINOVA,*[a] SARAH GILL,[b]
LAURA ST. ONGE[b] AND RICHARD L. SOWELL[a]

[a] WellStar College of Health & Human Services, Kennesaw State University,
Bldg. HHS 41, 1000 Chastain Road, Kennesaw, GA, USA; [b] KSU Club
Sports Owls Nest, Kennesaw State University, Bldg. 79, 1000 Chastain
Road, Kennesaw, GA, USA
*Email: sdambino@kennesaw.edu

8.1 Introduction

Traumatic brain injury (TBI) is a major health problem among young adults
ages 15 to 24 years. Sports and recreational activities contribute to approximately 21% of all cases of TBI.[1] Recent data suggests more than 62 000 cases
per year occur in contact sports in high school (i.e. ages 14 to 18 years).[2] 34% of
U.S. college football players have experienced at least one concussion and 20%
have experienced multiple concussions.[3] Sports-related concussion offers an
important research opportunity to investigate the utility of early biomarkers for
mild TBI because the time of head injury is known and blood samples can be
drawn within 24 hours.

Immediate, secondary and cumulative consequences follow mild TBI.
Cerebral edema and associated increased intracranial pressure are the major
immediate consequences of TBI.[4] Traumatized brains also have increased sensitivity to secondary ischemic insult, which is triggered by a neurotransmitter

RSC Drug Discovery Series No. 24
Biomarkers for Traumatic Brain Injury
Edited by Svetlana A. Dambinova, Ronald L. Hayes and Kevin K. W. Wang
© Royal Society of Chemistry 2012
Published by the Royal Society of Chemistry, www.rsc.org

storm evoked by TBI.[4] The cumulative incidence of mild TBI may result in post-traumatic epilepsy (PTE), which occurs in 4.4 per 100 persons with mild TBI in the first 3 years after hospital discharge.[5]

Acute subclinical concussions may be associated with microscopic widespread axonal injury and loss of neuronal connectivity (temporary denervation) which may initiate minor ischemic or hemorrhagic complications due to compression or decompression of microvessels in nervous tissue.[6] Normally, this subtle cell impairment is reversible and not visible on conventional CT and MRI images. Detection of impairment requires advanced radiological methods to recognize such abnormalities in grey or white matter. Only the world's most powerful magnetic resonance imaging machine, the 9.4 Tesla, is able to observe metabolic processes in a single neuron.[7] However, the 9.4 Tesla is currently only used in clinical research and is not yet available for routine assessments, particularly in the emergency department (ED) setting.

Diagnosis of concussion is complicated because many primary concussions go unrecognized or are not reported. This situation is especially true when a concussion is related to a sport injury, where loss of consciousness is rare. Additionally, competitive athletes are often subjected to recurrent concussions. If left unrecognized, such concussions may lead to potentially more debilitating delayed second-impact injuries.

Currently available blood-borne biomarkers that indicate brain injury, S-100,[8] NSE,[9] and cleaved Tau protein,[8] are still in the research phase and have not yet demonstrated satisfactory sensitivity or specificity in recognizing asymptomatic cases of concussions.[10] The excitatory neurotransmitter receptors as biomarkers of neurotoxicity for acute and chronic conditions of brain injury represent a promising avenue for selection of preventive treatment and follow up.[11] The latter raises three significant questions: (i) for injured athletes, is prolonged resting after a concussion adequate to overcome consequences of multiple impacts?; (ii) should neuroprotective treatment be administered to "perfectly" healthy young athletes?; and (iii) if so, how do you determine the necessity of such treatment?

In cases of uncertainty as to existence of a concussion, additional investigations combining results from advanced MRI, neurotoxicity biomarker assays, and neurocognitive testing may be required to improve diagnostic certainty of acute and chronic concussions. In this chapter, we present recently obtained data concerning the possibility of post-concussion effect assessment in club sport athletes with semi-acute and chronic conditions.

8.2 Biomarkers for Concussions

Concussion or minor impact head injury is a brief episode of focal and/or diffuse neurological dysfunction that involves post-concussive symptoms and altered mental status.[10] Confusion, post-traumatic amnesia, visual problems, ringing in the ears, and headache are common signs after acute concussions.

However, some cases, especially primary ones, may not have any clinical signs or develop symptoms for several days or weeks.[10]

The immediate damage due to severe or moderate TBI is recognizable and can be confirmed by brain imaging; namely, conventional CT scan. However, when diagnosing concussions or mild TBI, CT and MRI scans are often negative. There is no objective evidence that mild impact leads to structural damage; however, it can trigger metabolic and functional disturbances, causing development of long-term secondary impairment in nervous tissue. There is a critical need to develop a method or test for early recognition of the brain metabolic and functional disturbances and to predict/prevent worsening of these abnormalities. The goal of such a test would be to aid in selection and prompt administration of neuroprotective therapy that will delay or reverse secondary damage after the neurotrauma.

Approximately one third of the brain is devoted to the mechanics of vision and visual processing.[12] Many concussions, especially coup-contrecoup injuries, can result in focal or diffuse axonal injuries that affect temporary memory loss, migraine, and abnormal spiking activity on EEG.

Minor impact stimulates a molecular chain reaction of neurotoxicity in which long-term depolarization and dysfunction in signal transduction evokes brief neuronal firing.[12] Ionotropic receptors (AMPA, NMDA, kainite) cause an opening of channels, allowing potassium to flow out and calcium and sodium to enter,[13] leading to risk of post-traumatic headache or abnormal spiking activity (see Chapter 4).[14,15] A study to examine the presence of abnormal amounts of neurotoxicity biomarkers related to concussions was implemented at Kennesaw State University (KSU).

8.3 Neurotoxicity Biomarkers in Sport-Related Concussions

8.3.1 Healthy Volunteers and Club Sport Athlete Characteristics

The study protocol was reviewed and approved by the KSU Institutional Review Board. A total of 84 club sport athletes (56 male, 28 female; ages 20.8 ± 1.8 years) and 40 volunteers (non-athletes, 21 male, 19 female; ages 22.0 ± 4.1 years) were enrolled in the study between September and October 2011. Participating athletes represented the majority of the KSU Sports & Recreation Department, where rugby, soccer, lacrosse, and cheerleading occupied about 75% of the athletic curricula (see Figure 8.1). The average number of completed years of education for all participants was 13.4 ± 1.1.

Among the athletic participants, 47 (56%) reported having a collegiate level experience in a corresponding sport ranging from 0 to 2 years. Of this group of participants, 71% ($n = 33$) responded as having "no concussions" in neurocognitive testing. Additionally, a total of 51 athletes (60.5%) from all levels of athletic experience self-reported an absence of concussions. Of 84 athletes, 20 (23.7%) reported having had a single concussion, with the nearest impact injury

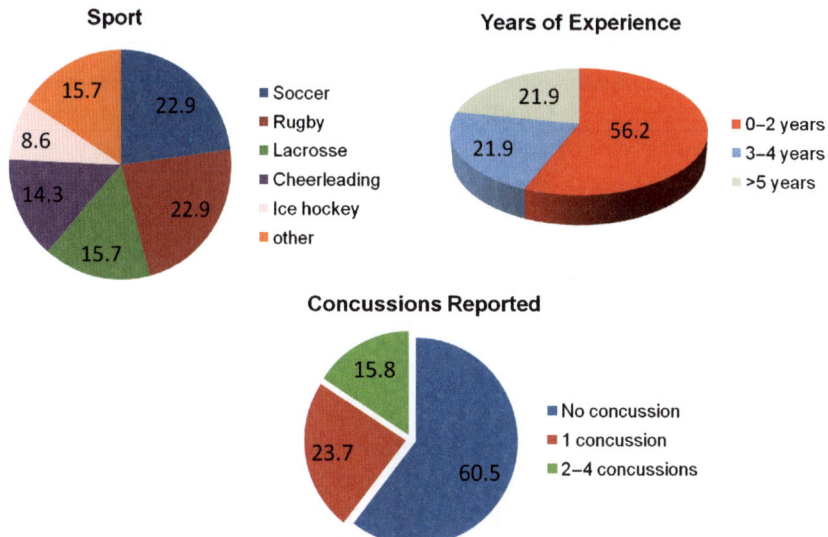

Figure 8.1   Information related to sport, experience, and concussions self-reported by KSU club sport athletes (56 male, 28 female), aged 21.0 ± 1.9 years. The circle of "Sports" represents the distribution of athletes ($n = 84$) in 14 of 24 clubs at KSU that participated in the study. The "Years of Experience" reflects the percentage of collegiate level of sport experience. Of 84 athletes, 51 (60.5%) from all levels of athletic experience self-reported the absence of concussions, 20 (23.7%) had a single concussion, and the remainder suffered 2–4 concussions, as presented in "Concussions Reported."

being 2 weeks prior to study enrollment. The remainder of athletes described a history of concussions, i.e. as having multiple (2–4) concussions, with a gap in time between recurrent concussions of more than 6 months.

There were five athletes with self-reported diagnosis of ADD, one of which additionally had learning problems defined by a history of past special education service.

8.3.2   Neurocognitive Testing and Neuroimaging

All study participants had baseline ImPACT testing (version 2.1), which showed all four average neurocognitive composite scores above the 30th percentile for age-normative parameters (see Table 8.1). These scores were significantly higher than proposed cutoff values for the four potential predictors at 80% sensitivity.[16] The athlete group scores did not differ from that of non-athlete controls, who had no history of concussions, indicating good recovery after the last impact for athletes.

Additionally, the post-concussion scale embedded in ImPACT testing was offered to all participants at baseline to mark from one to up to 22 commonly reported symptoms after concussions.[16] Baseline assessment showed that the

Table 8.1   ImPACT testing results (baseline) taken by KSU volunteers.

Parameter	Cutoff*	ImPACT result§	
		Non-athletes n = 40	Athletes n = 84
Verbal memory	64.5	90.0	88.9
Visual memory	46	77.5	76.5
Processing speed	23.5	40.1	41.0
Reaction time**	0.78	0.55	0.60
Cognitive efficiency	–	0.40	0.38

*At 80% sensitivity (Lau *et al.*, *Neurosurgery*, 2011, Aug 9, Epub ahead of print).
**For this variable, a higher score represented poorer performance, for remaining variables, a lower
    score represented worse performance.
§Data presented as mean.

athlete group, independently of number of concussions and the time since last injury, complained about poor memory and trouble concentrating, problems with falling asleep and drowsiness, consistent fatigue, and headaches.

Previous research[17] has found that mild TBI patients had significantly worse performance on working memory tasks in semi-acute (<2 weeks and 1 month) and chronic phases (1 year post injury) compared to matched controls. There was no significant correlation of 3T magnetic resonance imaging (3T MRI) findings with memory impairment.[17] However, 3T MRI scans detected parenchymal lesions in 75% of the mild TBI patient cohort with loss of consciousness and post-traumatic amnesia.[17] This represents a much higher rate than conventional 1.5 T MRI (23%).[18] Abnormalities in white matter of mild TBI patients at 24 hours and 1 month post injury were detected by diffusion tensor imaging (DTI), a relatively new MRI modality that capitalizes on the diffusion of water molecules for brain imaging.[19] Compared with traditional neuropsychological testing and conventional imaging techniques, DTI measures were more accurate in differentiating mild TBI from controls.[20] The functional MRI measured response after stimulus inception with processing speed (seconds to minutes) in mild TBI patients that cannot be captured by traditional neurocognitive testing.[21] It was suggested that functional imaging integrated with DTI may be more suitable to detect neuropsychological differences, including cognitive processing and outcomes from mild TBI.[22]

8.3.3   Reference Values of Neurotoxicity Biomarker Assays

AMPAR peptide values in plasma of non-athlete controls (no history of concussions) were detected by MP-ELISA[23] as 0.05–0.40 ng/mL for both genders (95%). About 5.0% of controls had AMPAR peptide levels of >0.40 ng/mL (see Table 8.2). The reference interval for GluR1 (AMPA receptor fragment) antibodies detected in serum by ELISA and calculated from the samples (85th percentile) was found to be 0.1–1.5 ng/mL and supported by results obtained earlier.[24] About 15% of the control cohort showed AMPAR peptide levels of >1.5 ng/mL.

Table 8.2 Neurotoxicity biomarkers reference values.

| Biomarker, ng/mL | Healthy individuals (n = 40) | | |
	N	% absolute	% population
AMPAR peptide			
<0.1	7	17.5	17.5
0.1–0.2	13	32.5	50.0
0.2–0.3	18	45.0	95.0
>0.4	2	5.0	100
GluR1 antibodies			
<0.5	2	5.0	5.0
0.5–1.0	21	52.5	57.5
1.0–1.5	11	27.5	85.0
>1.5	6	15.0	100
NR2 peptide			
<0.1	25	63.0	63.0
0.1–0.2	5	13.0	76.0
0.2–0.4	8	20.0	96.0
>0.5	2	4.0	100
NR2 antibodies			
<1.0	5	12.5	12.5
0.5–1.0	16	40.0	52.5
1.0–1.5	14	35.0	87.5
>1.5	5	12.5	100

NR2 peptide (NMDA receptor fragment) in plasma (MP-ELISA assay) and antibodies in serum (ELISA assay) of controls yielded in reference intervals of 0.01–0.5 ng/mL (96th percentile) and 0.5–2.0 ng/mL, respectively (see Table 8.2). Both NMDAR biomarker reference intervals were similar to previously published results.[25,26]

8.3.4 AMPAR Peptide/Antibodies

Of the 84 club sport athletes, 41 (49%) had higher AMPAR peptide concentrations (1.0–8.4 ng/mL) compared to that of healthy controls ($n = 40$, 0.4 ± 0.1 ng/mL, see Figure 8.2A). Abnormal AMPAR peptide concentrations did not depend on number of concussions and were detected in about 60% of freshmen athletes who did not self-report concussion. However, a comparison of the five scores of ImPACT testing for athletes with high AMPAR peptide levels and the non-athlete control group showed a tendency to decreased average scores for athletes (see Table 8.3). It included both memory components, processing speed, and reaction time; most interesting, a fifth component, cognitive efficiency, which has not previously been considered in other studies.[16,27]

GluR1 antibodies detected by ELISA were significantly elevated in the serum of three athletes compared with healthy non-athletes (<1.5 ng/mL) and accompanied drastically increased AMPAR peptide levels (see Figure 8.2B). Considering our previous experience with GluR1 antibody assay

Table 8.3   ImPACT testing parameters and AMPAR peptide.

ImPACT parameter	Controls $n = 40$	AMPAR Peptide $> 2.0\,ng/mL$ $n = 22$
Verbal memory	90.0	86.6
Visual memory	77.5	73.3
Processing speed	40.1	41.0
Reaction time*	0.55	0.6
Cognitive efficiency	0.40	0.34

*For this variable, a higher score represented poorer performance; for the remaining variables, a lower score represented worse performance.
All scores presented as average number.

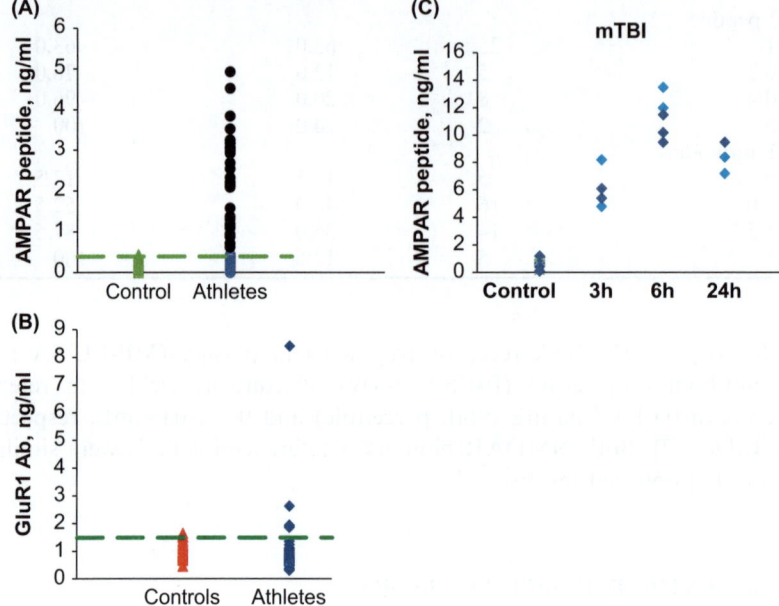

Figure 8.2   AMPAR peptide (A) and antibodies (B) in controls without history of concussions (non-athletes), $n = 40$, and athletes, $n = 84$. Dotted line shows cutoffs of 0.4 ng/mL and 1.5 ng/mL for AMPAR peptide and antibodies respectively. C: distribution of AMPAR peptide in plasma of athletes with concussions depending on time of admission after the impact injury.

usage,[23,24,28,29] these athletes would be recommended for MRI and an elec-troencephalographic (EEG) study to investigate possible encephalopathy and presence of abnormal brain spiking activity.

A previous pilot study related to sport and accident-related acute phase of mild TBI (age 18 ± 2.0 years, 7 male, 5 female) within 24 hours of brain injury (versus healthy controls, $n = 17$) examined at DeKalb Medical Center (Decatur, GA, USA) demonstrated increased AMPAR peptide as early as 3 hours post injury (see Figure 8.2C). No abnormal concentrations of GluR1 antibodies were detected in any of the patients with mild TBI in acute phase post injury.

Thus, the results of both studies described above indicate an opportunity for assessment of risk of encephalopathy in acute and chronic post injury in athletes prone to multiple concussions, as well as persons admitted to the ED with mild TBI symptoms.

8.3.5 NR2 Peptide/Antibodies

Figure 8.3 shows the distribution of plasma concentrations of NR2 peptide and serum content of NR2 antibodies in controls and athletes with concussions. The comparison of mean values of NR2 peptide/antibodies in independent age- and gender-matched groups demonstrated that values for the control represented low distributions of NR2 peptide and antibodies. Conversely, significant differences were observed for the concussion group. Of the 84 athletes enrolled in the study, 22 (26%) had increased NR2 peptide and 50 (60%) showed increased NR2 antibodies compared to the level for controls.

No changes in any of the five score ImPACT testing based on either the NR2 peptide or the NR2 antibody concentrations for athletes or controls were

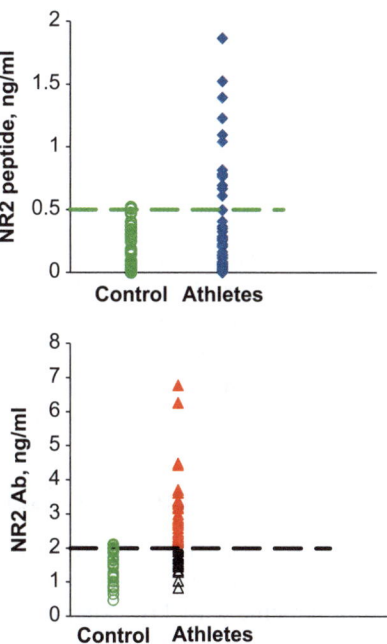

Figure 8.3 NR2 peptide (top) and antibodies (bottom) in controls without history of concussions (non-athletes), $n = 40$, and athletes, $n = 84$. Dotted line shows cutoffs of 0.5 ng/mL and 2.0 ng/mL for NR2 peptide and antibodies, respectively.

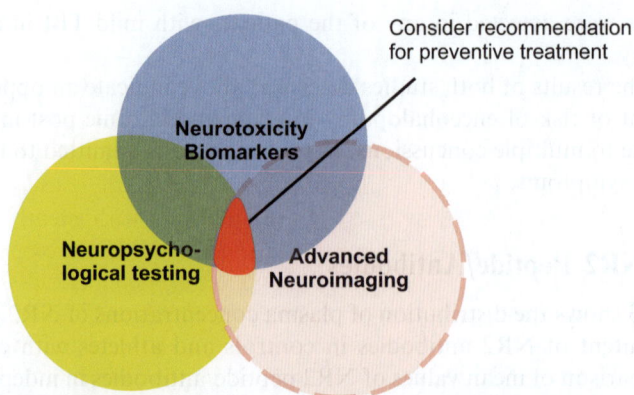

Figure 8.4 Post-concussion assessment for possible encephalopathy and cytotoxic edema is represented by the three circles. If an abnormal blood assay results coincides with advanced neuroimaging (DTI) data indicating axonal injury and decreased scores in neurocognitive testing (red triangle) for a person suspected of having a concussion, then the attending physician should consider preventive treatment.

registered. However, drastically increased NMDAR biomarker values indicated the possibility of cytotoxic edema due to secondary injury in acute and chronic phases post injury. This finding is supported by DTI findings of cytotoxic edema in semi-acute phase of mild TBI.[20]

In our current study of semi-acute to chronic phases of concussions in club sport athletes, the usage of biomarkers assays and ImPACT testing was useful for assignment for the neuroimaging study. Out of 84 observed athletes, 7 were recommended for MRI. In general, it seems plausible to consider a "three circles" approach to assessment of short- and long-term consequences after concussions and mild TBI in otherwise healthy persons (see Figure 8.4). The "three circles" approach comprises detection of neurotoxicity biomarkers in conjunction with advanced neuroimaging (3T or better 9T MRI, DTI, and functional MRI) and neuropsychological testing. Results could direct assignment of selected athletes to preventive treatment directed at improving encephalopathy and avoiding risk of neurological complications.

8.4   Risk Assessment of Secondary Ischemic Events

Concussion increases the risk of ischemic or hemorrhagic changes due to vascular system impairment, which may trigger a stroke.[30] During a 5-year follow-up clinical study, 8.2% of patients with TBI experienced stroke. The combined approach to early prognosis of possible stroke implying advanced neuroimaging, panel of biomarkers including NMDAR peptide and antibody assays, and neuropsychological testing could explain undetermined etiology of cryptogenic strokes in young persons.[31] This approach would be beneficial in preventing secondary deteriorating neurotoxic damage.

Specific neurotoxicity biomarkers might also assist in predicting secondary ischemic events after concussion/mild TBI. Studies have demonstrated that use of the NR2 peptide assay helped to distinguish TIA and ischemic stroke from stroke mimics in acute conditions,[25] while detection of NR2 antibodies was statistically predictive for stroke in patients with a history of diabetes, hypertension, or atrial fibrillation.[26] Additionally, results of studies using NR2 antibodies have revealed several asymptomatic cases with silently progressing ischemia. Other researchers have measured the anti-NMDA receptor antibody to distinct epitopes of the N-terminal domain in paraneoplastic disorders[32] and lupus erythematosus.[33] In both cases, about 6% of patients had a risk of stroke. To date, research that can detect undefined or under-recognized microscopic ischemic damages surrounding embolic teratoma or autoimmune vessel reactions by advanced neuroimaging remains in its infancy.

8.5 Risk Assessment of Cerebral Hematoma and Hemorrhage

Brain imaging is the gold standard for diagnosis of subarachnoid and intracerebral hemorrhages following TBI.[34] Utilizing blood tests in the pre-hospital and emergency management of mild TBI remains challenging. Rapid management of acute brain or spinal cord injuries in the ED setting, especially if the CT scan is normal or MRI is contraindicated or not available, is essential to ensure that patients receive proper treatment within the therapeutic window. The ability to speed up ruling in or ruling out axonal/neuronal or vascular damages prior to CT imaging using a blood test to detect a specific brain biomarker – or biomarkers – would be extremely useful.

One brain biomarker under study is glial fibrillary acidic protein (GFAP), a structural protein that maintains astroglial cell contact within the blood-brain barrier.[35] GFAP is released rapidly into the bloodstream when microscopic hematoma/hemorrhage leads to gradual necrosis and cytolysis of neuronal proteins related to the neurotoxicity cascade (see Chapter 10).

A multicenter trial was conducted to evaluate diagnostic accuracy of the NR2 peptide and GFAP biomarkers for rapid differentiation of hyperacute cerebral ischemia versus intracerebral hemorrhage (i.e. within 0.5 to 4.5 hours of onset).[36] In this 186-patient cohort, prevalence of intracerebral hemorrhage by CT was 25%. The NR2 peptide ruled in hyperacute ischemic stroke and revealed intracerebral hemorrhage ($\sim 11\%$) accompanied by ischemic complications, while GFAP recognized only cases of cortical hemorrhage. Use of the NR2 peptide and GFAP biomarkers may provide reliable assessment of intracerebral hemorrhage across a broad, clinically important range of patients with hyperacute hemispheric stroke, may improve diagnostic certainty of cerebral ischemia, and reveal intracerebral hemorrhage accompanied with ischemic complications. These two promising biomarker assays would also warrant study for their ability to assess mild TBI in conjunction with advanced radiological methods and neurocognitive testing.

8.6   Risk of Abnormal Cerebral Spiking Activity

Post-traumatic headache often can occur following mild TBI and usually resolves within months. However, in some cases, chronic headaches and abnormal brain spiking activity may result in long-term neurological complications. EEG studies in children with chronic post-traumatic headaches as a result of multiple concussions have shown nonspecific paroxysmal changes in 18% and epileptiform activity in 6%.[29] Simultaneous detection of neurotoxicity biomarkers in serum samples of these patients demonstrated a 150% increase in AMPA but not NMDA receptor antibodies. The highest levels of NMDA receptor antibodies were observed within the first year after injury in those with post-traumatic headaches due to multiple contusions.[29]

Clinical studies in patients with acute mild closed head injuries have found that 14% had increased NR2 peptide and developed symptoms of neurological deficit (e.g. temporary visual loss, sensory loss, confusion, or memory problems). Patients (45%) presenting with high levels of GluR1 peptide showed abnormal spiking activity on EEG; however, only those with abnormally high levels of both NR2 and GluR1 peptides developed brain-related seizures within 3 years of injury.[37]

Another consequence of mild TBI is increased subsequent risk (by approximately 50%) of post-traumatic seizures and epilepsy.[38] As a potentially serious secondary injury, post-traumatic seizures occur frequently in patients with TBI and in patients with traumatic parenchymal hemorrhage on admission to intensive care units.[39] Post-traumatic seizures are often associated with evidences of secondary excitotoxicity and precede to long-term abnormal brain spiking activity.

Early investigations have established the capability of antibodies to GluR2/3 to enhance spiking activity in epilepsy.[40,41] Antibodies to the GluR1 subunit of AMPA receptors were also observed in patients with post-traumatic epilepsy[42,43] (see Chapter 9).

8.7   Conclusions

Biomarkers of neurotoxicity may assist in the assessment of subtle, or asymptomatic, ischemic, hemorrhagic, or axonal injuries associated with abnormal spiking activity. Timely assessment of concussions is essential for optimal management of athletes and prevention of neurological complications. Clinical use of blood tests that can reliably predict consequences, or forecast recovery and outcome, might help in selection of the preventive therapy for improving acute or subacute health outcomes.[44]

Yet, there is no reliable, rapid, and affordable diagnostic tool that can speed up assessment of mild TBI and assist in determining when an injured athlete should return to play. The data presented here demonstrate that specific brain biomarkers detected in human biological fluids used in conjunction with advanced neuroimaging and neuropsychological assessment might improve diagnostic certainty and help facilitate diagnosis of mild TBI. These results could help guide the clinician (neurologist or neurosurgeon) or athletic trainer in preventing future neurological complications.

Rapid assessment of subclinical central nervous system injuries using appropriate blood tests would be useful in various clinical settings, e.g. emergency departments and primary care offices as well as in-patient and out-patient facilities. Such tests could monitor treatment dosage and outcome. The latter would have the potential to shorten the often lengthy recovery process after head and spinal cord injuries and subsequent neurological complications. These results would not only benefit the patient, but would be an important advantage to the healthcare system by reducing direct and indirect costs.

References

1. M. Faul, L. Xu, M. M. Wald and V. G. Coronado, *Traumatic Brain Injury in the United States: Emergency department visits, hospitalizations and deaths 2002–2006*, Atlanta, GA, CDC, 2010, available at www.cdc.gov/ ncipc/tbi/tbi.htm
2. J. W. Powell and K. D. Barber-Foss, *JAMA*, 1999, **282**, 958.
3. E. F. Luckstead, *AAP Grand Rounds*, 2004, **11**, 16.
4. K. K. Jain, *Drug Discovery Today*, 2008, **13**, 1082.
5. P. L. Ferguson, G. M. Smith, B. B. Wannamaker, D. J. Thurman, E. E. Pickelsimer and A. W. Selassie, *Epilepsia*, 2010, **51**(5), 891.
6. T. W. McAllister, in *Traumatic Brain Injury*, 2nd edn., ed. J. M. Silver, T. W. McAllister and S. C. Yudofsky, American Psychiatric Publishing, Inc., Arlington, VA, 2011, p. 239.
7. I. C. Atkinson, A. Lu and K. R. Thulborn, *Magn. Reson. Med.*, 2011, **66**(4), 1089.
8. J. J. Bazarian, F. P. Zemlan, S. Mookerjee and T. Stigbrand, *Brain Inj.*, 2006, **20**(7), 759.
9. C. Geyer, A. Ulrich, G. Gräfe, B. Stach and H. Till, *J. Neurosurg. Pediatr.*, 2009, **4**, 339.
10. G. A. Davis, G. L. Iverson, K. M. Guskiewicz, A. Ptito and K. M. Johnston, *Br. J. Sports Med.*, 2009, **43**(Suppl I), i36.
11. R. L. Blaylock and J. Maroon, *Surg. Neurol. Inter.*, 2011, **2**, 107.
12. E. R. Kandel, T. M. Jessell and J. R. Sanes, in *Principles of Neural Science*, 4th edn., ed. E. R. Kandel, J. H. Schwartz and T. M. Jessell, McGraw-Hill, New York, 2000, p. 533.
13. C. Betzen, R. White, C. M. Zehendner, E. Pietrowski, B. Bender, H. J. Luhmann and R. W. Kuhlmann, *Free Radical Biol. Med.*, 2009, **47**, 1212.
14. B. Yasseen, A. Colantonio and G. Ratcliff, *Brain Inj.*, 2008, **22**, 752.
15. T. D. Seifert and R. W. Evans, *Curr. Pain Headache Rep.*, 2010, **14**, 292.
16. B. C. Lau, M. W. Collins and M. R. Lovell, *Neurosurgery*, 2011, Aug 9, Epub ahead of print.
17. H. Lee, M. Wintermark, A. Gean, J. Ghajar, G. T. Manley and P. Mukherjee, *J. Neurotrauma*, 2008, **25**, 1049.

18. E. Kurca, S. Sivak and P. Kucera, *Neuroradiology*, 2006, **48**, 661.
19. H. G. Belanger, R. D. Vanderploeg, G. Curtiss and D. L. Warden, *J. Neuropsychiatry Clin. Neurosci.*, 2007, **19**(1), 5.
20. A. R. Mayer, J. Ling, M. V. Mannell, C. Gasparovic, J. P. Phillips, D. Doezema, R. Reichard and R. A. Yeo, *Neurology*, 2010, **74**, 643.
21. A. R. Mayer, M. V. Mannell, J. Ling, R. Elgie, C. Gasparovic, J. P. Phillips, D. Doezema and R. A. Yeo, *Hum. Brain Mapp.*, 2009, **30**(12), 4152.
22. E. D. Bigler and J. J. Bazarian, *Neurology*, 2010, **74**, 626.
23. A. Shikuev, A. Bagumyan, U. Danilenko and S. Dambinova, *J. Neurotrauma*, 2011, **28**, A-55.
24. S. A. Dambinova, G. A. Izykenova, S. V. Burov, E. V. Grigorenko and S. A. Gromov, *J. Neurol. Sci.*, 1997, **152**, 93.
25. S. A. Dambinova, *Clin. Lab. Inter.*, 2008, **32**, 7.
26. J. D. Weissman, G. A. Khunteev, R. Heath and S. A. Dambinova, *J. Neurol. Sci.*, 2011, **300**(1–2), 97.
27. P. Schatz, J. E. Pardini, M. R. Lovell, M. W. Collins and K. Podell, *Arch. Clin. Neuropsychol.*, 2006, **21**, 91.
28. A. Shikuev, T. Skoromets, G. Izykenova, U. Danilenko and S. Dambinova, *AACC Annual Meeting Abstracts*, 2011, A137.
29. A. V. Goryunova, N. A. Bazarnaya, E. G. Sorokina, N. Yu. Semenova, O. V. Globa, Zh. B. Semenova, V. G. Pinelis, L. M. Roshal and O. I. Maslova, *Neurosci. Behav. Physiol.*, 2007, **37**, 761.
30. Y. H. Chen, J. H. Kang and H. C. Lin, *Stroke*, 2011, **42**(10), 2733.
31. P. E. Cotter, P. J. Martin and M. Belham, *Intern. J. Stroke*, 2011, **6**(5), 445.
32. P. Niehusmann, J. Dalmau, C. Rudlowski, A. Vincent, C. E. Elger, J. E. Rossi and C. G. Bien, *Arch. Neurol.*, 2009, **66**, 458.
33. L. A. DeGiorgio, K. N. Knostantinov, S. C. Lee, J. A. Hardin, B. T. Volpe and B. Diamond, *Nat. Med.*, 2001, **7**(11), 1189.
34. C. M. Micheel and J. R. Ball, in *Evaluation of Biomarkers and Surrogate Endpoints in Chronic Disease*, The National Academy, Washington DC, 2010, p. 22.
35. H. Reiber, *Restor. Neurol. Neurosci.*, 2003, **21**, 79.
36. C. Foerch, M. Niessner, T. Back, M. Bauerle, G. M. De Marchis, A. Ferbert, H. Grehl, G. F. Hamann, A. Jacobs, A. Kastrup, S. Klimpe, F. Palm, G. Thomalla, H. Worthmann and M. Sitzer for the BE FAST Study Group, *Clin. Chem.*, 2012, **58**, 237.
37. S. A. Dambinova, *IVD Technology*, 2007, **3**, 15.
38. J. A. Langlois, A. Marr, J. Mitchko and R. L. Johnson, *J. Head Trauma Rehabil.*, 2005, **20**, 196.
39. P. M. Vespa, D. L. McArthur, Y. Xu, M. Eliseo, M. Etchepare, I. Dinov, J. Alger, T. P. Glenn and D. Hovda, *Neurology*, 2011, **75**, 792.
40. S. W. Rogers, P. I. Andrews, L. C. Gahring, T. Whisenand, K. Cauley, B. Crain, T. E. Hughes, S. F. Heinemann and J. O. McNamara, *Science*, 1994, **265**, 648.

41. R. E. Twyman, L. C. Gahring, J. Spiess and S. W. Rogers, *Neuron*, 1995, **14**, 755.

42. M. M. Odinak, D. E. Dyskin, I. A. Toropov, A. Y. Emelin, A. A. Cherepanov and S. A. Dambinova, *Zh. Nevrol. Psikhiatr. Im. S. S. Korsakova*, 1996, **96**, 45.

43. S. A. Gromov, S. K. Khorshev, Yu. I. Poliakov, L. G. Gromova and S. A. Dambinova, *Zh. Nevropatol. Psikhiat.r Im. S. S. Korsakova*, 1997, **97**, 46.

44. N. Bailey, in *Nutrition and Traumatic Brain Injury*, ed. J. Erdman, M. Oria and L. Pillsbury, The National Academies Press, Washington, DC, 2011, p. 315.

CHAPTER 9

Feasibility Studies of Neurotoxicity Biomarkers for Assessment of Traumatic Brain Injury

ALEXEY V. SHIKUEV,*[a] TARAS A. SKOROMETS,[a] DMITRI I. SKULYABIN,[b] MIROSLAV M. ODINAK[b] AND ALEXANDER A. SKOROMETS[a]

[a] Pavlov State Medical University, St. Petersburg, Russia; [b] Russian Medical Military Academy, St. Petersburg, Russia
*Email: shikerman@mail.ru

9.1 Introduction

Mild traumatic brain injury (TBI) is the most prevalent form of head injury in civilian and military settings. Diagnosis of mild TBI poses some difficulties, particularly the challenges of determining consequences after the injury. As a multi-factorial condition, TBI is manifested throughout the continuum of care, causing a risk of development in the future of post-traumatic epilepsy (PTE), stroke, and other neurological conditions.[1,2]

Accurate and early identification of concussions and mild TBI using brain biomarker assays in conjunction with advanced neuroimaging has the potential to reduce the number of undiagnosed cases, which, if left untreated, can lead to more debilitating and even fatal second-impact injuries.[3] Such confirmation may also assist physicians in determining readiness of a person to return to duty (physical activity, military service, and sports play) after a concussion.

RSC Drug Discovery Series No. 24
Biomarkers for Traumatic Brain Injury
Edited by Svetlana A. Dambinova, Ronald L. Hayes and Kevin K. W. Wang
© Royal Society of Chemistry 2012
Published by the Royal Society of Chemistry, www.rsc.org

This approach might help triage persons seeking immediate neuroprotective treatment to prevent secondary injuries.[4]

Neurotoxicity biomarker blood tests may also have the potential to stratify the risk of possible consequences after concussions and mild TBI. These assays may serve as prognostic tools to assess the degree of brain injury causing epileptiform activity or/and cerebrovascular accident, or stroke. The latter could have a significant impact on patient care by assisting in diagnosis and management of patients with brain injuries (see Chapter 4).

This chapter is devoted to clinical feasibility studies of: (i) AMPAR peptide in assessment of patients with mild TBI; (ii) AMPAR antibodies in evaluation of PTE in persons after moderate TBI; and (iii) predicting risk of stroke after mild TBI using NMDAR peptide and antibodies.

9.2 Biomarkers of Chronic Encephalopathy

Mild TBI is the net effect of chronic encephalopathy that may be temporary (primary) concussions, long lasting (secondary), or even result in a permanent disruption of neuronal connectivity in one or more regions of the brain.[5] In addition to impaired oxidative metabolism,[6] the post-traumatic neurometabolic cascade may include hyperglycolysis, accumulation of lactate, enzyme-activated apoptosis, disrupted cytoskeletal architecture, axon swelling and secondary axotomy, free radical production and inflammation, impaired connectivity and altered neurotransmission, and altered cerebral blood flow[7] (see Figure 9.1).

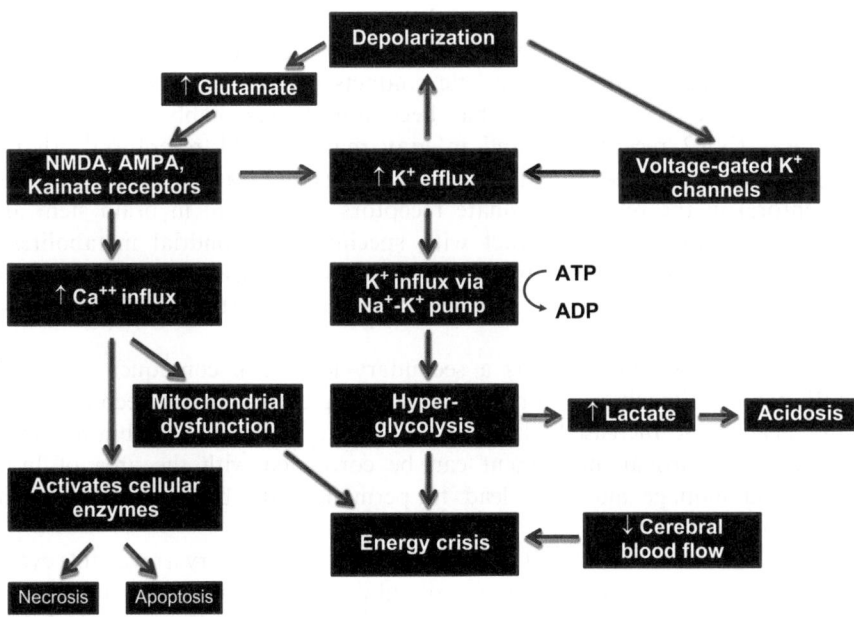

Figure 9.1 Metabolic cascade of mild TBI.

Table 9.1   Cell apoptosis and necrosis.

Pathological features	*Apoptosis*	*Necrosis*
Patterns of death	Single cells	Groups of neighboring cells
Cell size	Shrinkage, fragmentation	Swelling
Plasma membrane	Preserved continuity Phosphatidylserine on surface	Smoothing, early lysis
Mitochondria	Increased membrane permeability Structure relatively preserved	Swelling, disordered structure
Organelle shape	Contracted "Apoptotic bodies"	Swelling, disruption
Nuclei	Chromatin: clumps and fragmented	Membrane disruption
DNA degradation	Fragmented internucleosomal cleavage free 3′ ends	Diffuse and random
Cell degradation	Phagocytosis No inflammation	Inflammation, macrophage invasion

These structural and metabolic abnormalities may be detected by advanced radiological methods (diffusion tensor imaging, functional MRI, and DWI).[8]

Many secondary pathological events underlie multiple mild TBI initiated by energy failure and protein synthesis/degradation processes (see Chapter 4). Following injury, blood flow is altered,[9,10] resulting in neurotoxicity (excitotoxicity) that leads to cell death through apoptosis or necrosis (Table 9.1). It is known that AMPA receptors are mostly situated in axonal and dendritic contacts of the frontal lobe, hippocampus, amygdala, and cerebellum. In cerebellar Purkinje cells, which normally lack the NMDA receptor,[11] the majority of synaptic contacts contain AMPA/kainite receptors. NMDA receptor density has been found presumably on microvessel surfaces (NR2 receptors)[12] and in gray matter (NR1 receptors)[13] that is engaged in microvessel function and synaptic signal transduction (NR1/NR2 receptors) in the retina.[14] Kainate receptors are present in brain stem and spinal cord axons and interact with specific mitochondrial metabolites.[15] Therefore, ionotropic glutamate receptors may be associated with certain locations of the injury and related to specific subcortical/axonal symptom presentation.[6]

Chronic encephalopathy, as a secondary long-term consequence of mild TBI, confers difficulties in concentration during daily activity, sleep problems, and headaches. Increased likelihood of stroke, epilepsy, and brain tumors caused by neuronal impairment can be correlated with the area of brain structural damage and may lead to permanent disabilities with enduring neurotoxic damage.[2,6]

Early diagnosis of mild TBI is critical to prevent secondary irreversible events occurring in the brain. A diagnosis of subtle damages will allow tailoring of timely, appropriate treatment to prevent neurotoxic damage, which can otherwise lead to a number of neurological conditions.[2,5,16]

9.3 Mild TBI in Trauma Unit Setting

Approximately 1.0 million Russians each year are involved in violence-related events, motor vehicle crashes, and incidental falls that result in a TBI. TBI occurs in 400 to 720 of 100 000 persons annually; mild TBI accounts for approximately 81% to 90% of all impacts.[17] We used blood assays in conjunction with CT/MRI to validate these four neurotoxicity biomarkers – AMPAR peptide/GluR1 antibodies and NR2 peptide/antibodies – in the assessment of patients with mild TBI admitted to a trauma unit.

9.3.1 Human Subject Characteristics

After approval by the institutional review board, a cohort of patients with TBI admitted to the Trauma Unit at A. L. Polenov Neurosurgical Institute (St. Petersburg, Russia) from June 2010 through October 2011 was recruited. A total of 155 patients and controls (88 male, 67 female) were included in the study. Mean age was 36.5 years (SD 9.1, range 19–56 years). All patients were healthy prior to the TBI ($n = 80$) with no history of neurological or psychiatric disorders. The gender- and age-matched control group comprised 75 subjects, including healthy persons without a history of TBI, neurological or psychiatric disorders, and drug use ($n = 28$); patients with transient ischemic attack (TIA, $n = 15$) within 12 hours of symptom onset; and patients with spinal cord ischemia ($n = 28$) or stroke ($n = 4$).

TBI patients presented within 5 days of brain injury due to criminal activity (49% physical assault), moving vehicle accident (30%), and incidental fall (21%). Thirty one suffered from mild TBI according to a Glasgow Coma Scale (GCS) score of > 13 upon admission (in 85%, the documented score was 15). There were 30 persons with moderate TBI (GCS score of 10–12), three patients with severe TBI (GCS of 7–8), and 16 with polytrauma.

Demographic and clinical information collected included age, gender, type of injury (TBI or polytrauma), history of head injury, and severity of TBI based on GCS. Neurological examination, including MRI and CT scans, were performed for all TBI patients.

CT head scans undertaken in observed TBI subjects ($n = 80$) were normal for those with mild TBI; in moderate and severe cases, areas of parenchymal and intracerebral hematoma/hemorrhage were registered. MRI depicted cytotoxic edema (3–5 mL) in 42% of patients with mild TBI ($n = 13$). Areas that were predominantly affected by impact included the frontal/anterior temporal lobe cortex (43%) and, less commonly affected, the lateral midbrain (22%), the inferior cerebellum, and the midline superior cerebral cortex (11%). In our subset of patients who suffered mild TBI without loss of consciousness ($n = 17$), basal ganglia hypoperfusion abnormalities were registered in 45% patients as second to frontal lobe abnormalities and more common than temporal lobe abnormalities. Neuroimaging (diffusion weighted imaging) in 7 persons with mild TBI recorded multiple small lesions and edemas in the frontal cortical areas of the patients.

9.3.2 AMPAR Peptide and Antibody

In this study, a selective increase in AMPAR peptide values was observed in patients with semi-acute mild TBI within 5 days of the impact (see Figure 9.2A). Four patients with mild TBI had AMPAR peptide concentration below 0.4 ng/mL, which may have been due to areas of microscopic hemorrhage formed after the impact that could not be detected by this biomarker.

We did not observe a correlation of AMPAR peptide with severity of TBI in this feasibility study. Low AMPAR peptide levels detected in the blood of patients with moderate and severe TBI probably connected with bigger size of hemorrhages (see Chapter 2) and impaired blood circulation. In this case, the peptide has limited access to bloodstream due to clotting and the surge of pro-teases activity that may digest AMPAR fragments. The same effect was observed for NR2 peptide in large hemispheric strokes (>25 mL), even though no cor-relation of biomarker concentrations with volume of ischemic lesions observed.[20]

To clarify the critical cut-off value for AMPAR peptide concentrations with the best performance for mild TBI, three distinct control groups comprising healthy volunteers, persons with spinal cord ischemia, and stroke-like symp-toms in addition to severe TBI and polytrauma were studied. The cut-off value of 0.4 ng/mL was yielded from data calculations (see Figure 9.2A). AMPAR peptide values increased above the cut-off were detected in two patients with moderate TBI (0.5–1.10 ng/mL) and four patients with spinal cord ischemia (0.64–3.47 ng/mL). Clinical symptoms were clearly present in these patients with moderate TBI (GCS scores of 13) but negative images on standard CT were registered, indicating that such cases should be investigated further utilizing advanced neuroimaging.

The analysis of tradeoffs between true-positive and false-positive rates for the AMPAR peptide assay is shown by presenting data as a traditional Receiver Operating Curve (ROC, see Figure 9.2B). The proportional area under the curve was 0.885. The sensitivity of 84% and specificity of 93%, corresponding to a cut-off of 0.4 ng/mL with a significant positive likelihood ratio of 11.6 to diagnose mild TBI, have been calculated. Measurement of AMPAR peptide with a cut-off value of 0.4 ng/mL in the emergency department may have two potential clinical indications: (a) rule in a patient with concussions and mild TBI and (b) rule out other neurological complications or polytrauma.

Measurement of GluR1 antibodies (antibodies to GluR1 subunit of AMPA receptors) in blood samples of enrolled patients (see Figure 9.2C) showed increased antibodies in one case each in the mild and moderate TBI groups (about 3.5 ng/mL); in one patient with polytrauma (2.2 ng/mL); and in three patients with spinal cord ischemia (3.1–9.7 ng/mL). Abnormal amounts of GluR1 antibodies detected in serum of patients usually indicates an increased risk for abnormal brain spiking activity and development of "chronic" conditions.[18,19]

Both assays may be used for ruling out individuals without mild TBI. If the test at a cut-off point of 0.4 ng/mL were negative, the post-test probability for mild TBI would be $< 4\%$. The latter could be an approach to speed up ruling out of mild TBI and select patients who should be immediately directed to

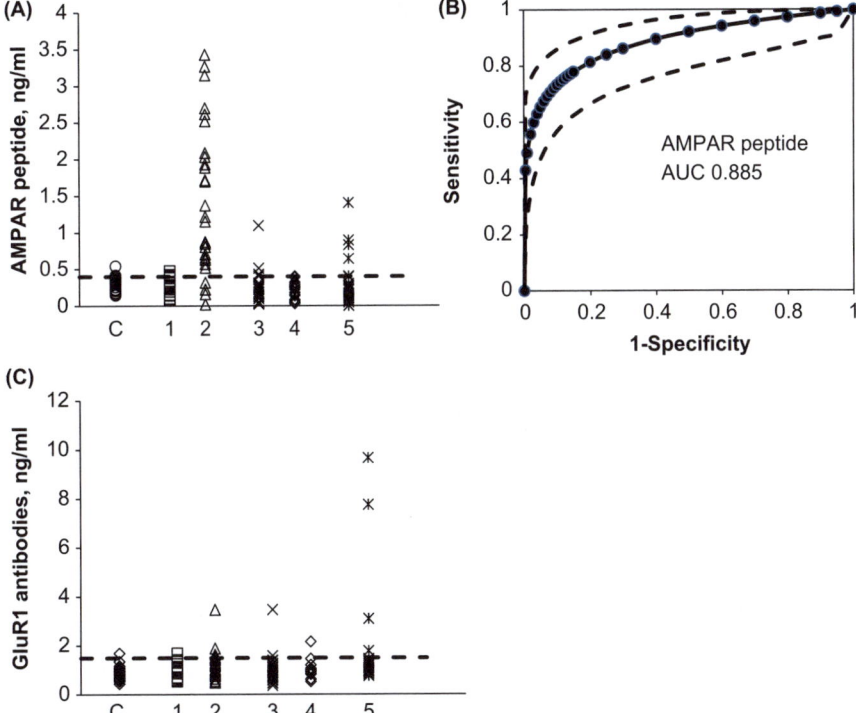

Figure 9.2 AMPAR biomarkers in patients admitted to the emergency department. A: AMPAR peptide detected in plasma of healthy controls (C, $n = 28$), patients with neurological disorders including TIA (1, $n = 19$), mild TBI (2, $n = 31$), moderate TBI (3, $n = 30$), severe TBI and polytrauma (4, $n = 19$), spinal cord ischemia (5, $n = 28$). Dashed line shows cut off for AMPAR peptide assay (0.4 ng/ml). B: fitted receiver operating curve (ROC) demonstrated the performance of plasma AMPAR peptide assay to assess mild TBI versus other groups (moderate and severe TBI, polytrauma, and other neurological disorders). Dashed lines indicate 95% confidence interval of the fitted ROC curve. Area under the curve (AUC) is 0.885. C: GluR1 antibodies measured in serum samples from patients (same indications). Dashed line shows cut off of 1.5 ng/ml for GluR1 antibody assay.

neuroimaging. It may also optimize use of neuroimaging in the most cost effective manner possible.[21]

9.3.3 NR2 Peptide and Antibody

Once aware of mild TBI, we assessed priority and risk categorization of possible stroke(s) due to compression of brain microvessels. Figure 9.3A shows the distribution of plasma concentrations of NR2 peptide and serum content of NR2 antibodies in controls and those with mild TBI. The comparison of mean values of NR2 peptide in independent age- and gender-matched groups demonstrated that values for the controls belonged to low distributions.

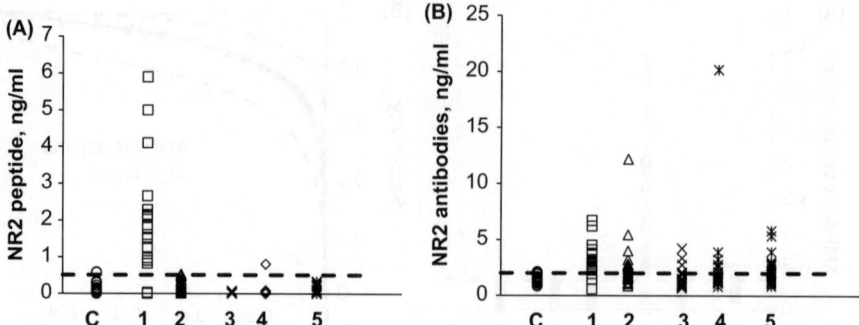

Figure 9.3 NMDAR biomarkers in patients admitted to ED. A: NR2 peptide
detected in plasma of healthy controls (C, $n = 28$), patients with neurological disorders including TIA (1, $n = 19$), mild TBI (2, $n = 31$), moderate
TBI (3, $n = 30$), severe TBI and polytrauma (4, $n = 19$), spinal cord ischemia
(5, $n = 28$). B: NR2 antibodies measured in serum samples from patients
(same indications). Dashed lines show cut offs for NR2 peptide (0.5 ng/ml)
and NR2 (2.0 ng/ml) respectively.

Significant differences in NR2 peptide concentrations were observed only for
patients with TIA and stroke compared to all TBI and spinal cord ischemia
groups (Figure 9.3A). As expected, most patients with TIA and stroke enrolled
within 12 hours of symptoms onset showed high values of NR2 peptide that
were above the 0.5 ng/mL cut-off earlier determined for patients in the emergency department.[22] The opportunity to speed up ruling in or ruling out of
ischemic versus non-ischemic events might be critical to expedite timely treatment. The latter may help to reduce the number of ischemic complications as a
long-term consequence after mild TBI.

NR2 antibodies measurements in serum samples of 155 patients showed
values that were increased above the 2.0 ng/mL cut-off[23] in representatives of
every group compared to a healthy population (see Figure 9.3B). NR2 antibodies are associated with risk of cerebral ischemic events.[22] Depending on the
level of NR2 antibodies measured in the blood, the likelihood of secondary
cerebral ischemic event(s) may increase significantly after mild TBI.

9.4   Mild TBI in the Military Setting

Brain injuries caused by explosions have become some of the most common
combat wounds suffered in Iraq. According to the U.S. Department of Defense,
the Military Health System has recorded 43 779 patients who have been diagnosed with a TBI in 2003 through 2007.[24] The multi-center Defense and Veterans
Brain Injury Center has reported treating 2669 patients during this period;
however, physicians believe many less obvious brain injury cases go undetected.[25]

Mild TBI (primarily loss of consciousness) was reported to occur in 22% of
all brain injuries after the Second Chechnya War.[26] Cumulative incidence of

mild TBI may result in post-traumatic epilepsy (PTE), which occurs in 4.4 per 100 persons with mild TBI in the first 3 years after hospital discharge.[1] Soldiers sometimes walk away from explosions with no obvious injuries, and they can recover with rest and time away from the battlefield. But the military estimates that one-fifth of troops with these mild injuries will have prolonged or life-long symptoms requiring continuing care.

The goal of this study was investigate clinical feasibility of AMPAR peptide/ antibodies biomarkers to assess mild TBI in military personnel as an aid to neuroimaging and cognitive testing.

9.4.1 Clinical and Neuroimaging Findings

The study protocol was approved by the Medical Military Academy Ethics Board (St. Petersburg, Russia). 173 subjects (123 male, 50 female) with recurrent mild TBI due to blast injury (1 week after injury) were included in the study. Mean age of the subjects was 23 years (23.4 ± 4.2 years; range, 19–25 years). These subjects with mild TBI were identified as a part of post-deployment TBI screening at the Medical Military Academy (St. Petersburg, Russia), and all of the subjects received standard CT scans. Those with mild TBI presented with GCS scores of 13–15; 112 patients (65%) with mild TBI had a score of 15; 45 patients (26%), a score of 14; and 16 patients (9%) presented with a GCS score of 13. These subjects with mild TBI all had witnessed loss of consciousness of less than 30 minutes and presence of post-traumatic amnesia as part of the study inclusion criteria. They did not report any other neurological or psychiatric diagnosis, including substance or alcohol abuse.

Clinical neuroimaging findings were normal in the majority of mild TBI cases. Of the CT examinations in all 173 subjects with mild TBI, intraparenchymal lesions were detected in 17 CT scans (9.8%) (see Figure 9.4). Abnormal CT findings were noted in 5 (4.5%) soldiers with mild TBI and a GCS score of 15; in 8 (18%) with a GCS score of 14; and in 4 (25%) with a GCS score of 13. Abnormalities detected in CT scans were similar to those reported previously in many other clinical studies.[27,28] However, the cumulative incidence of intraparenchymal lesions (9.8%) in our study in a military population was lower than CT pathological findings (50%) reported earlier in civilian individuals.[29]

The gender- and age-matched control group comprised subjects ($n = 64$, aged 22.4 ± 4.2 years; 35 male, 31 female) with no history of head injury with persisting symptoms or complaints, no central neurological disorder or psychiatric condition, and no regular intake of psychoactive drugs or history of drug abuse. CT scans of all volunteers were negative.

9.4.2 Cognitive Evaluation

To evaluate cognitive functions after mild TBI, a Mini Mental State Examination (MMSE) was completed on admission as 15-minute battery of tests by all participants.[30] While the MMSE has limited specificity with respect to individual clinical syndromes, it represents a brief, standardized method by

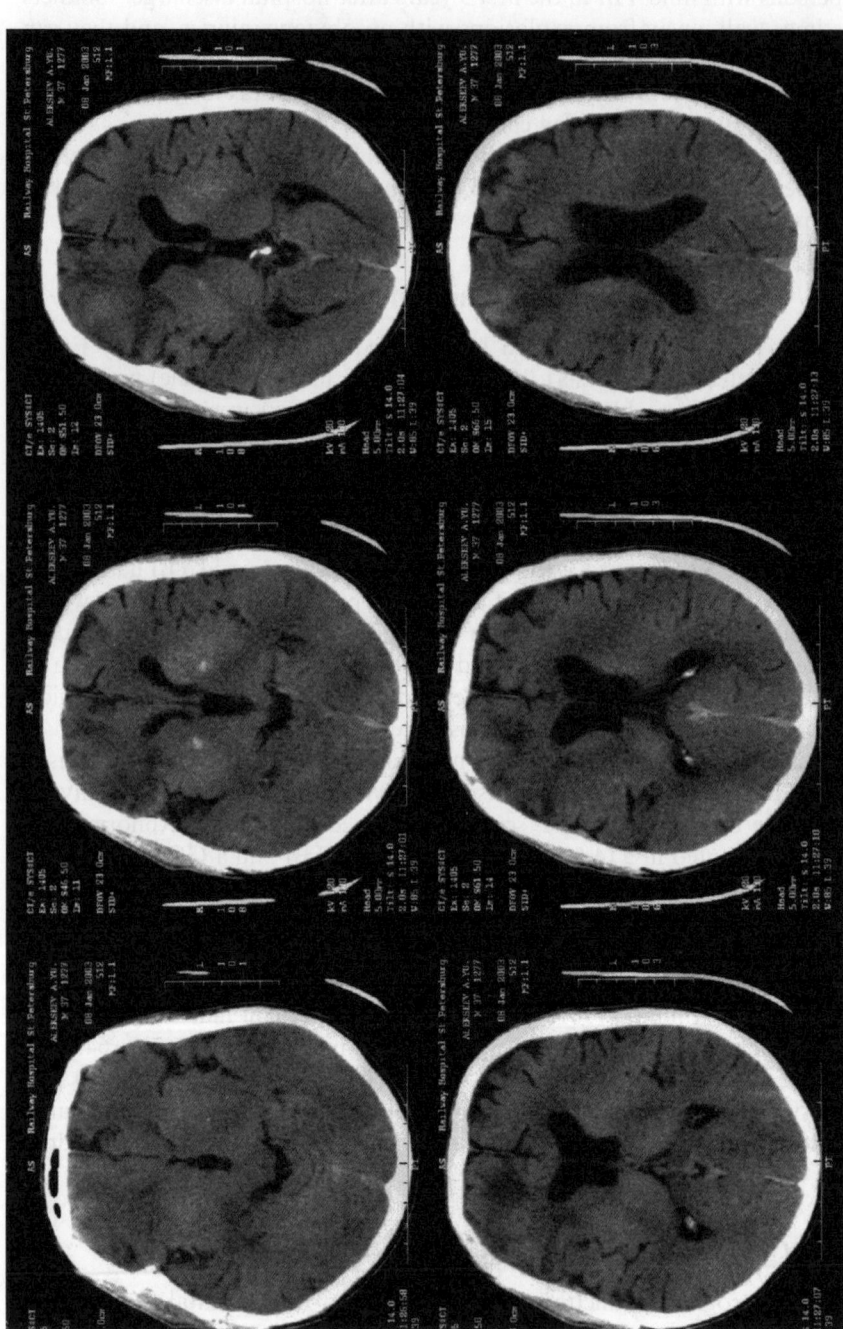

Figure 9.4 CT scans of a 26-year-old male active duty soldier with mild TBI after exposure to an explosive blast during combat operations. CT performed on 7th day after brain injury.

which to grade cognitive mental status. It assesses orientation, attention, immediate and short-term recall, language, and the ability to follow simple verbal and written commands. Furthermore, it provides a total score that places the individual on a scale of cognitive function.

Evaluation of cognitive impairment in adults recognized a high prevalence of mental dysfunction in individuals who suffered mild TBI. Mean baseline of total MMSE results (28.4 ± 2.3 points) were significantly decreased in patients with mild TBI (23.8 ± 2.6 points, $P < 0.05$); in particular, the attention, recall, and orientation component scores (Figure 9.5A,B). Three MMSE components (attention, orientation, and recall) were significantly affected by mild TBI. Data of impaired cognitive functions obtained from active duty personnel with mild TBI did not contradict results of an earlier study and demonstrated typical neuropsychological consequences of mild TBI: reduced attention, slowing of information processing, and worsening of memory and learning abilities.[31]

9.4.3 AMPAR Peptide in Plasma of Active Duty Personnel

Distributions of AMPAR peptide values in plasma samples from apparently healthy males ($n = 35$) and apparently healthy females ($n = 31$) in the clinically relevant age range of 19–26 years were evaluated (Table 9.2). The reference interval calculated from the samples (central 86th percentile) was found to be 0.1–1.0 ng/mL for both genders. Approximately 14.1% of the apparently healthy population showed AMPAR peptide levels > 1.0 ng/mL.

AMPAR peptide concentrations in plasma for each group are shown in Figure 9.5C, where 121 individuals with mild TBI had an increased peptide levels with an average concentration of 2.98 ng/mL (range, 2.0–6.1 ng/mL). 14 patients with recurrent mild TBI had lower AMPAR peptide concentrations, with average of 0.91 ng/mL (range, 0.75–1.00 ng/mL). Healthy controls had a mean value of the peptide of 0.49 ng/mL (range, 0.01–1.22 ng/mL) (see Figure 9.5C). The intra-assay coefficient of variation (CV) was 5.3% to 6.3%, and the inter-assay CV was 5.9% to 9.8%.

The optimal cut-off value for recurrent mild TBI was 1.0 ng/mL (92% sensitivity, 81% specificity), at which a positive predictive value of 93% was achieved. The tradeoffs between true-positive and false-positive rates are shown by presenting data as a traditional ROC curve (see Figure 9.5D). The proportional area under the curve was 0.97.

Table 9.2 AMPAR peptide reference values.

| AMPAR peptide, ng/mL | Healthy individuals ($n = 64$) | | |
	N	% absolute	% population
<0.1	19	29.7	29.7
0.1–0.5	23	35.9	65.6
0.6–1.0	13	20.3	85.9
>1.0	9	14.1	100

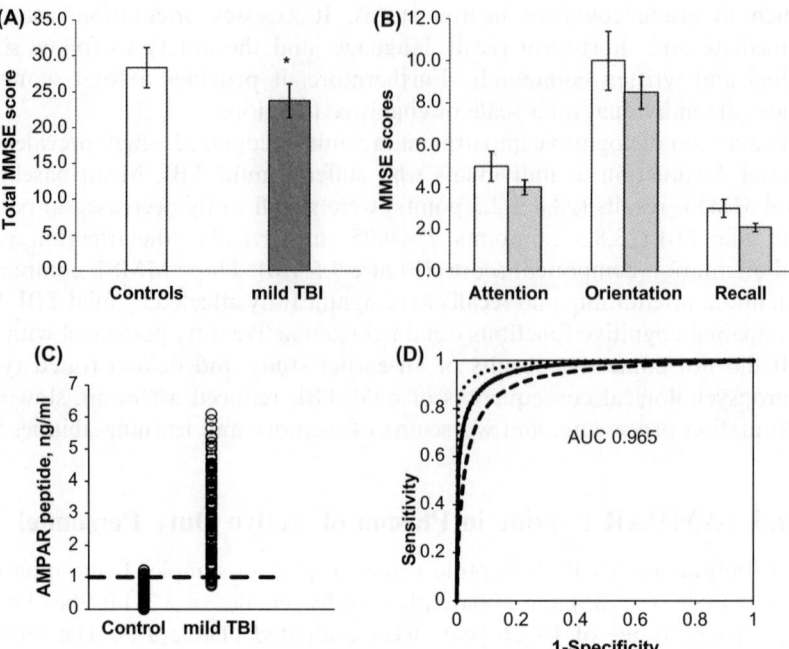

Figure 9.5   A: mean baseline total MMSE scores for healthy individuals ($n = 64$) and
active duty personnel with mild TBI ($n = 173$). MMSE component scores
B: attention, orientation, and recall in control group of healthy persons
(white bars) and patients with neurotrauma (dark bars). C: AMPAR
peptide levels in plasma of healthy controls and active duty personnel with
mild TBI. D: fitted receiver operating curve (ROC) demonstrating the
performance of plasma AMPAR peptide assay to assess mild TBI versus
controls. Dashed lines indicate 95% confidence interval of the fitted ROC
curve. Area under the curve (AUC) is 0.965.

It is necessary to note that military personnel with mild TBI who had positive
and negative CT scans showed increased AMPAR peptide levels. However, in
individuals with positive CT findings ($n = 17$), AMPAR peptide concentrations
were higher (range, 4.1–6.1 ng/mL) compared to those with mild TBI, who had
negative CT findings (range, 1.2–4.5 ng/mL).

Thus, in case of recurrent mild TBI, the AMPAR peptide assay cut-off of
1.0 ng/mL might serve as a potential threshold for ruling out other conditions,
including but not limited to healthy controls, concussions, and most cases of
moderate and severe TBI described in section 9.3.

9.5   Post-Traumatic Epilepsy in Active Duty Personnel Following Moderate TBI

As reported earlier, the incidence rate of post-traumatic seizures is about 22%.[9]
Seizures may occur immediately following the trauma, although post-traumatic

epilepsy (PTE) usually develops within months to year after the injury. While immediate post-traumatic seizures may be successfully treated, the best estimate of the effect of anti-epileptics is a reduction in seizures of $<25\%$.[32]

This study explored the association of AMPAR biomarkers with the extent of brain structural and functional abnormalities in patients with post-traumatic epilepsy who had undergone mild to moderate TBI 1 year previously and in those with temporal lobe epilepsy (TLE) diagnosed within the past 1 to 3 years.

9.5.1   Study Participants

Patients with PTE who were followed for 1 year after mild to moderate TBI were recruited from the Neurology Hospital Medical Military Academy (St. Petersburg, Russia) between 2000 and 2002. 44 patients (28 male and 16 female; 35.4 ± 7.2 years) were diagnosed as suffering from PTE with partial seizures; in 21 patients, secondary generalization was reported. A diagnosis of PTE was established in patients after mild to moderate TBI who had at least one epileptic seizure. 93 patients with TLE (44 male and 49 female; 28.1 ± 5.2 years) causing partial ($n = 78$) and tonic-clonic ($n = 15$) seizures within the previous 1 to 3 years were included in the study.

The non-epileptic control group ($n = 61$) consisted of subjects with no history of central neurological disorders or psychiatric conditions and no regular intake of psychoactive drugs or history of drug abuse. There were 42 males and 19 females in the control group with mean age of 28.4 ± 6.0 years. The local ethics committee approved the project and written consent was obtained from all participants.

All patients had a neurological examination, including assessment of cranial nerves, motor and sensory systems, deep tendon reflexes, and coordination; an electroencephalogram (EEG); and an MRI scan.

9.5.2   Anatomical and Functional Assessments
of Epileptiform Activity

In this study, we investigated anatomical and functional characteristics of symptomatic patients following mild to moderate TBI and TLE. The pattern of traumatic lesions on MRI images has been found in 78.5% patients with PTE and located in the right and left frontal lobe, occipital lobe, lateral temporal lobe, and basal ganglia. Only one patient (3%) with PTE demonstrated isolated hippocampal sclerosis that was characterized by selective neuronal loss and gliosis in hippocampal subfields CA1, Ca3, and dentate gyrus.[33] No pathological changes on MRI images were found in 7 patients with PTE. An MRI study of those suffering TLE demonstrated isolated hippocampal sclerosis with 10% to 25% atrophy in 51 patients (54.8%). The loss of more than 25% hippocampal volume in 27 patients (29%) with TLE was observed.

EEG with abnormal epileptiform activity was detected in 75% of patients with PTE and abnormal neuronal EEG activity in 55.8% of those with TLE. A vast majority of seizure patterns were observed over the lesioned lobe areas in patients with PTE. Mean duration of an individual seizure was between 2 and 3 minutes. The seizures frequently were clustered and occurred over a several hour period prior to stopping. Seizures were focal in origin and had secondary generalization in 47.7% of cases.

9.5.3 AMPAR Antibodies in Serum of Persons with Post-Traumatic Epilepsy and Temporal Lobe Epilepsy

Detection of AMPAR peptide in both group of patients with PTE and TLE showed approximately the same levels and were comparable with that of healthy controls. Due to the chronic nature of the patients' conditions (followed 1 year after injury), detection of abnormally high AMPAR peptide levels, reflecting acuteness of mild TBI, was not anticipated.

GluR1 antibodies measured in serum of patients revealed significantly increased amounts in patients with PTE ($P < 0.001$) and TLE ($P < 0.01$) compared to non-epileptic controls (see Figure 9.6). Patients with PTE had 2.5 ± 0.34 ng/mL (range, 1.64–3.12 ng/mL) level of GluR1 antibodies while for patients with TLE somewhat lower levels of antibodies of 1.96 ± 0.29 ng/mL (range, 1.26–2.56 ng/mL) were calculated.

Within each group of patients, the highest GluR1 antibody levels were observed in those with partial seizures, especially with secondary generalization, supporting results reported previously.[34] Abnormal concentrations of antibodies were detected in all patients with PTE, including in seven without structural alterations on MRI images but who demonstrated epileptiform activity on EEG and were diagnosed as status epilepticus. In patients with hippocampal atrophy, slightly reduced GluR1 antibodies that were still higher than that of the control group might be explained by gliosis of CA1, CA3, and

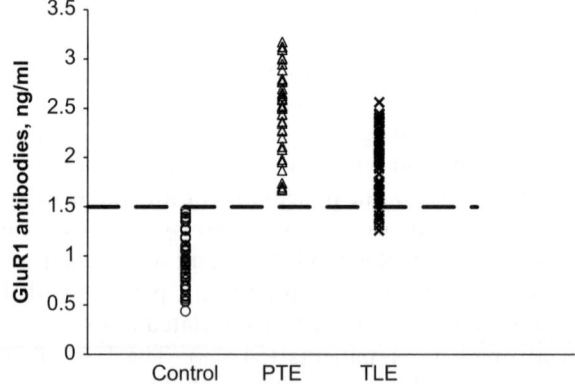

Figure 9.6   The distribution of GluR1 antibodies in healthy controls ($n = 61$), patients with post-traumatic epilepsy (PTE, $n = 44$), and patients with temporal lobe epilepsy (TLE, $n = 93$).

dentate gyrus structures, where the GluR1 subunit of AMPA receptors is located in abundant amounts under normal conditions.[35] Intractable seizures, often present in patients with TLE, cause degradation of GluR1 receptors[36] and their drain[18] through an already disrupted blood-brain barrier.[37]

9.6 Conclusions

Given that brain pathology underlying TBI is very complex in that it leads to various mental and neurological deficits, several biomarkers are likely to be needed to diagnose neurotraumatic sequelae. We propose a panel of biomarkers that includes four neurotoxicity markers reflecting acute state and risk of chronic sequelae after mild TBI.

The clinical feasibility study demonstrated that AMPAR peptide has the potential to reveal acute and semi-acute mild brain injury with 84% to 92% sensitivity and 81% to 93% specificity, depending on the setting. The simultaneous abnormal increase in GluR1 antibodies is associated with risk of abnormal brain spiking activity. These biomarkers, when used in a panel, could assist in ruling in or ruling out possible seizures and recognize non-convulsive seizures after mild TBI. Further clinical studies needed to provide more information concerning the value of AMPAR peptide/antibodies as an aid to diagnosis of TBI and the severity of impact.

In addition, NR2 peptide and antibody assays have shown a tendency to reveal acute and chronic secondary ischemic events following mild to moderate TBI that directly impairs nervous tissue microvessels. An abnormal increase in at least three of the four neurotoxicity biomarkers in patients with TBI may well indicate status epilepticus with subsequent atrophy of some of the subcortical structures, including the hippocampus.

References

1. P. L. Ferguson, G. M. Smith, B. B. Wannamaker, D. J. Thurman, E. E. Pickelsimer and A. W. Selassie, *Epilepsia*, 2010, **51**(5), 891.
2. Y. H. Chen, J. H. Kang and H. C. Lin, *Stroke*, 2011, **42**(10), 2733.
3. I. Cernak and L. J. Noble-Nacusslein, *J. Cereb. Blood Flow Metab.*, 2010, **30**(2), 255.
4. K. K. Jain, *Drug Discov. Today*, 2008, **13**(23–24), 1082.
5. S. A. Dambinova, in *Therapeutic Electrical Stimulation of Human Brain and Nerves*, ed. N. P. Bechtereva, Sova, St. Petersburg, 2008, p. 346.
6. A. R. Zazulia, T. O. Videen and W. J. Powers, *Stroke*, 2009, **40**, 1638.
7. G. Barkhoudarian, D. A. Hovda and C. C. Giza, *Clin. Sports Med.*, 2011, **30**, 33.
8. G. A. Davis, G. L. Iverson, K. M. Guskiewicz, A. Ptito and K. M. Johnston, *Br. J. Sports Med.*, 2009, **43**(Suppl I), i36.
9. P. M. Vespa, *Stroke*, 2009, **40**, 1547.

10. A. I. Qureshi, Z. Ali, M. F. Suri, A. Shuaib, G. Baker, K. Todd, L. R. Guterman and L. N. Hopkins, *Crit. Care Med.*, 2003, **31**, 1482.
11. M. Renzi, M. Farrant and S. G. Cull-Candy, *J. Physiol.*, 2007, **585**, 91.
12. C. D. Sharp, M. Fowler, T. H. Jackson, J. Houghton, A. Warren and A. Nanda, *BMC Neurosci.*, 2003, **4**, 28.
13. G. G. Nagy, M. Watanabe, M. Fukaya and A. J. Todd, *Eur. J. Neurosci.*, 2004, **20**, 3301.
14. V. P. Connaughton, in *Glutamate Receptors in Peripheral Tissue: Excitatory Transmission Outside the CNS*, ed. G. Santokh and O. Pulido, Kluwer Academic/Plenum Publishers, New York, 2010, p. 99.
15. J. L. Rozas, in *Kainate Receptors: Novel Signaling Insights (Advances in Experimental Medicine and Biology)*, ed. A. Rodriguez-Moreno and T. S. Sihra, Springer, New York, 2011, p. 72.
16. N. R. Temkin, *Epilepsia*, 2009, **50**(Suppl 2), 10.
17. K. E. Poshataev, *Far East Medical J.*, 2010, **4**, 125 (in Russian).
18. S. A. Dambinova, G. A. Izykenova, S. V. Burov, E. V. Grigorenko and S. A. Gromov, *J. Neurol. Sci.*, 1997, **152**(1), 93.
19. S. A. Dambinova, O. K. Granstrem, A. Tourov, R. Salluzzo, F. Castello and G. A. Izykenova, *J. Neurochem.*, 1998, **71**, 2088.
20. S. A. Dambinova, G. A. Khounteev, G. A. Izykenova, I. G. Zavolokov, A. Y. Ilyukhina and A. A. Skoromets, *Clin. Chem.*, 2003, **49**, 1752.
21. J. Undén and B. Romner, *J. Head Trauma Rehabil.*, 2010, **25**(4), 228.
22. S. A. Dambinova, *Clin. Lab. Inter.*, 2008, **32**, 7.
23. J. D. Weissman, G. A. Khunteev, R. Heath and S. A. Dambinova, *J. Neurol. Sci.*, 2011, **300**(1–2), 97.
24. US Military Casualty Statistics, http://www.fas.org/sgp/crs/natsec/RS22452.pdf, 2009, Mar 25.
25. K. M. Hall, Army conducts brain tests on soldiers, *Associated Press Writer*, 2007, Sept., 19.
26. V. G. Pomnikov, E. V. Korchagina and S. A. Shahbanov, *Neuroimmunology*, 2007, **7**(2), 94 (in Russian).
27. J. Borg, L. Holm, J. D. Cassidy, P. M. Peloso, L. J. Carroll, H. von Holst and K. Ericson, *J. Rehabil. Med.*, 2004, **Suppl. 43**, 61.
28. H. G. Belanger, R. D. Vanderploeg, G. Curtiss and D. L. Warden, *J. Neuropsychiatry Clin. Neurosci.*, 2007, **19**(1), 5.
29. H. Lee, M. Wintermark, A. D. Gean, J. Ghajar, G. T. Manley and P. Mukherjee, *J. Neurotrauma*, 2008, **25**, 1049.
30. P. M. Bokesch, G. A. Izykenova, J. B. Justice, K. A. Easley and S. A. Dambinova, *Stroke*, 2006, **37**(6), 1432.
31. E. Kurca, S. Sivak and P. Kucera, *Neuroradiology*, 2006, **48**, 661.
32. N. R. Temkin, *Epilepsia*, 2009, **50**(2), 10.
33. R. P. Bote, L. Blazquez-Llorca, M. A. Fernandez-Gill, L. Alonso-Nanclares, A. Munoz and J. De Felipe, *Semin. Ultrasound CT MR*, 2008, **29**, 2.
34. S. A. Gromov, S. K. Khorshev, Yu. I. Poliakov, L. G. Gromova and S. A. Dambinova, *Zh. Nevropatol. Psikhiatr. Im. S. S. Korsakova*, 1997, **97**, 46 (in Russian).

35. N. C. Day, T. L. Williams, P. G. Ince, R. K. Kamboj, D. Lodge and P. J. Shaw, *Brain Res. Mol. Brain Res.*, 1995, **31**(1–2), 17.
36. E. V. Grigorenko, S. Glazier, W. Bell, M. Tytell, E. Nosel, T. Pons and S. Deadwyler, *Neurol. Sci.*, 1997, **153**, 35.
37. O. Tomkins, A. Feintuch, M. Banifla, A. Cohen, A. Friedman and I. Shelef, *Cardiovasc. Psychiatry Neurol.*, 2011, published online, Feb 22, 2011.

CHAPTER 10

Astroglial Proteins as Biomarkers of Intracerebral Hemorrhage

CHRISTIAN FOERCH

Department of Neurology, Goethe-University, Frankfurt am Main, Germany
Email: foerch@em.uni-frankfurt.de

10.1 Introduction

Metaphorically speaking, rapid expansion of spontaneous arterial bleeding into the brain parenchyma is a severe form of "traumatic" brain injury, leading to irreversible destruction of neurons, astroglial cells, and the intercellular matrix. On a molecular level, the pathophysiological alterations that emerge in the time course of traumatic brain injury and intracerebral hemorrhage (ICH), respectively, may be comparable to a certain degree, and biomarker candidates that emerge as diagnostic markers in one of these diseases may also qualify in the other.

10.2 Intracerebral Hemorrhage: Epidemiology and Pathophysiology

Worldwide, ICH accounts for 10% to 15% of all strokes. Incidence rates range from 10 to 20 cases per 100 000 in predominantly white populations to 50 cases per 100 000 in predominantly Asian populations and blacks.[1–5]

RSC Drug Discovery Series No. 24
Biomarkers for Traumatic Brain Injury
Edited by Svetlana A. Dambinova, Ronald L. Hayes and Kevin K. W. Wang
© Royal Society of Chemistry 2012
Published by the Royal Society of Chemistry, www.rsc.org

Compared with ischemic stroke, ICH is associated with higher mortality rates and worse long-term prognosis. Six months after symptom onset, 25% to 50% of all patients with acute spontaneous ICH will have died; many more remain severely disabled.[5–7]

Based on underlying pathophysiology, ICH can be classified as either primary or secondary.[5] Primary ICH comprises the spontaneous rupture of a small perforating brain artery (diameter 50–200 μm) associated with arterial hypertension. Hypertensive brain bleeds are most often located deep within the brain, including the basal ganglia, thalamus, brain stem, and cerebellum. Long-term arterial hypertension provokes degenerative alterations of the vessel wall, a process termed lipohyalinosis. Histologically, lipohyalinosis is characterized by subintimal deposition of lipid-filled macrophages, fibroblast proliferation, and the replacement of smooth muscle cells in the tunica media with collagen.[8] These changes reduce the compliance of the vessel wall, thereby increasing the susceptibility to a spontaneous rupture.[2,9] According to electron microscopy studies, bleeding typically occurs at or near the bifurcation of affected arterioles, where the degenerative changes are found to be most prominent.[10] The rupture of dilated parts of the arteriolar wall, or Charcot-Bouchard microaneurysm, was reported to be the definitive cause of bleeding.[9,11] Subsequently, these microaneurysms were found to be small spots of extravascular blood resulting from a previous vascular leakage that are surrounded by either remains of the vessel wall or fibrin.[12,13] Other authors reported formation of a dissecting aneurysm in the vessel wall after a primarily intimal lesion to be the determining cause of ICH.[12]

Cerebral amyloid angiopathy (CAA) is another major cause of primary ICH. Due to insufficient removal of amyloid beta proteins in the brain, layers of congophilic material are deposited in the tunica media and adventitia of cortical and meningeal vessels.[8,14,15] These amyloid accumulations are associated with a fibrinoid necrosis of the vessel wall and the formation of microaneurysms, resulting in a leakage of blood into the brain parenchyma. In contrast to hypertension-associated ICH, CAA patients more often bleed into the peripheral lobar white matter near the boundaries between the gray and white matter. Both large parenchymal bleeds and cerebral microhemorrhages can occur.

In contrast to primary ICH, secondary ICH typically results from rupture of preformed vascular malformations (such as dural fistula or pial arteriovenous shunts), but can also occur due to brain tumors or an impaired status of coagulation.[1,5]

10.3 Hematoma Expansion and Cellular Destruction

Both in primary and secondary ICH, rupture of an arterial blood vessel gives rise to rapidly expanding bleeding into the brain parenchyma. By means of a "domino effect", the growing hematoma provokes subsequent mechanical damage to surrounding blood vessels and induces secondary bleeds.[9] Systematic investigations based on repetitive brain imaging revealed that ongoing bleeding beyond the first few hours is a frequent finding that occurs in

11% to 38% of all cases of ICH, depending on time between symptom onset and first brain scan.[16,17]

Among factors that reduce and ultimately stop bleeding expansion, activation of the coagulation system is crucial. At the initial bleeding site itself as well as those of mechanically induced secondary hemorrhage, "bleeding globes" can be found, an accumulation of platelets with concentric lamellae of fibrin.[9] The role of the coagulation system in preventing excessive hematoma expansion becomes particularly evident in patients who develop ICH while on effective warfarin anticoagulation. These patients have larger hematoma volumes, increased hematoma growth, and a worse prognosis.[18,19] In a mouse model of collagenase-induced ICH, warfarin pretreatment more than doubled hematoma volume compared to mice with normal coagulation status.[20,21] Interestingly, antiplatelet agents appear to interfere less critically in this scenario. Both clinical and experimental data suggest that antiplatelet pretreatment does not lead to increased hematoma volumes and a worse prognosis.[22,23] Taken together, these observations underline the importance of the plasmatic coagulation system in terminating active intraparenchymal bleeding. Conversely, activation of the coagulation system in patients with normal coagulation status by means of intravenous injections of recombinant factor VII reduced hematoma growth, but did not improve functional outcome.[24] Hematoma expansion is also determined by blood pressure in the arterial system.[25] Recent data impressively show that rapid reduction of blood pressure in the hyperacute phase of ICH formation (i.e. within 6 hours of symptom onset) leads to smaller hematoma volumes and less hematoma growth.[26] A steady state between blood pressure in the ruptured arterial vessel on one side and the counter-pressure of the surrounding brain tissue on the other ("tamponade") may be necessary to terminate hematoma expansion in addition to blood coagulation.

On a cellular level, shear stress and mass effects during hematoma expansion cause rapid mechanical disruption of brain tissue and lead to instant necrosis of neuronal and glial cells as well as to destruction of the intercellular matrix and the blood-brain barrier (BBB).[5,27,28] In brains of animals subjected to collagenase-induced ICH, severe cellular destruction at the bleeding site can be detected as early as 1 hour after induction of hemorrhage (Figure 10.1A). This includes a "chaotic" necrosis with structural disintegration of cells and cytolysis. The intracellular contents, including cytoplasm and the nucleus, are released from cell bodies into the extracellular space. Histologically, this severe destruction of tissue in the very early phase is a key feature of acute ICH, which is essentially different from ischemic stroke, e.g. caused by thromboembolic occlusion of arterial brain supply. In ischemic stroke, signs of necrosis and cytolysis are not detectable until 6 to 12 hours after symptom onset (Figure 10.1B). In earlier phases of cerebral ischemia (e.g. within 3 hours of symptom onset), brain cells are characterized by condensated nuclei and swollen cell bodies, but the structural integrity of the cell is still preserved and the intracellular content is not yet released into the extracellular space.[29–32]

Following the first wave of mechanical destruction during the phase of hematoma expansion, multiple biochemical cascades are activated in the

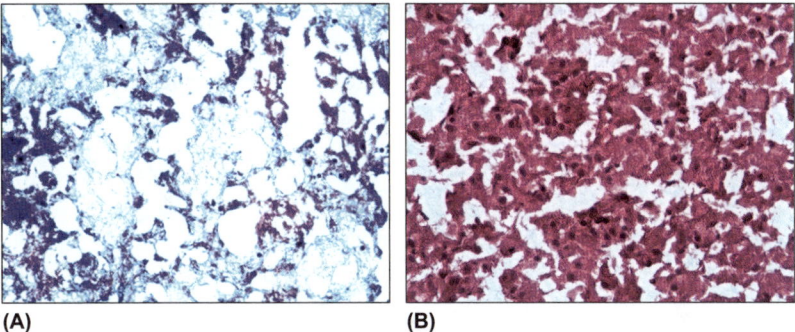

(A)   **(B)**

Figure 10.1   Panel A shows HE staining of a mouse brain (centered at the right striatum) 1 hour after induction of intracerebral hemorrhage (ICH) by means of collagenase injection into the right striatum. Note the destruction and structural disintegration of cells with release of the intracellular content and the nucleus into the extracellular space. Panel B shows the same brain area in a mouse that was subjected to a 1 hour filament occlusion of the middle cerebral artery. Although some cell swelling is apparent, the structural integrity of the cells is well preserved.

perihematoma region, inducing secondary brain injury. This includes formation of a predominantly vasogenic edema with additional mass effects and activation of matrix metalloproteinases with subsequent destruction of the BBB, as well as neurotoxic effects of hemoglobin degradation products and iron and induction of inflammatory processes after infiltration of immune cells, such as macrophages and neutrophils.[5,33,34] This secondary injury is considered to contribute independently to ICH prognosis, and is therefore a target of neuroprotective strategies.[35]

10.4   Astroglial Proteins as Biomarkers of Intracerebral Hemorrhage

Based on pathophysiological differences between ICH and ischemic stroke in terms of the kinetics of cellular disintegration and cytolysis, one may assume that any protein concentrated in brain cells in higher amounts than in the surrounding extracellular space and systemic circulation is released more rapidly in ICH than in ischemic stroke (Figure 10.2). Thus, the number of potential biomarkers for ICH is theoretically high. However, at present, sufficient data are available only for protein S100B and GFAP.

10.4.1   Protein S100B

For years, protein S100B has been known as one of the most abundantly present proteins in the brain. It is found in the cytoplasm and nucleus of astrocytes, oligodendrocytes, and Schwann cells. After calcium binding, S100B

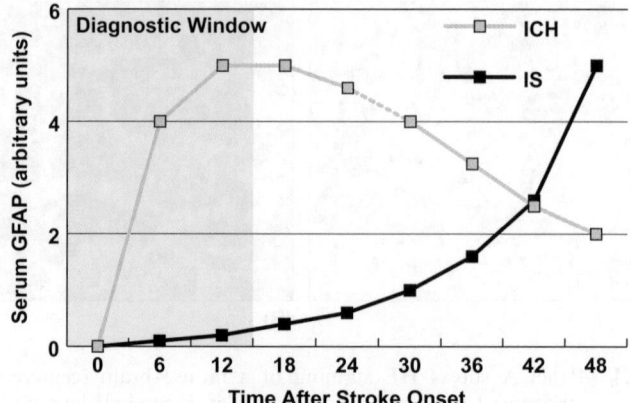

Figure 10.2 Hypothetical release kinetics of GFAP in patients with intracerebral hemorrhage (ICH, grey line) and ischemic stroke (IS, black line). Due to different kinetics of astroglial necrosis in these two subtypes of stroke, a diagnostic window exists within the first 6–12 hours after symptom onset, in which GFAP values are high in cases of ICH, but low in IS.

undergoes conformational changes, which allows this protein to interact with various biochemical pathways. Among others, S100B is involved in regulation of protein phosphorylation and cell–cell signaling.[36,37] In the 1960s, S100B was considered to be brain specific, but subsequently was also identified in extra-cerebral tissues, such as adipocytes and melanoma cells.[38,39] Protein S100B is detectable in the blood of healthy individuals in concentrations between 0.05 and 0.15 µg/L.[40] Thus, release of S100B from the brain into the blood following brain injury or stroke may potentially be masked by S100B found in the bloodstream under physiological conditions.

10.4.1.1 Protein S100B in the Blood of Patients with ICH

Several studies have analyzed serum levels of protein S100B in patients with acute ICH. Weglewski *et al.* found peak S100B concentrations on day 1 after ICH onset, whereas patients with ischemic stroke revealed a more delayed S100B release, reaching maximum concentrations around day 3.[41] Delgado *et al.* determined S100B concentrations in 78 ICH patients within 24 hours of symptom onset.[42] In comparison to healthy controls, median S100B values were found elevated in ICH patients. S100B levels correlated with ICH volume and were identified as a predictor of early neurological deterioration and a worse functional outcome.

Hu *et al.* investigated release kinetics of S100B into serum in patients with ICH in more detail, including 86 with acute ICH.[43] They found elevated S100B serum concentrations as early as 6 hours after symptom onset with peak levels at 24 hours. Higher S100B serum concentrations were associated with larger ICH volumes and reduced survival rates. In a multivariate model, the S100B serum concentration at hospital admission was identified to be an independent predictor of short-term mortality. Specifically, a S100B serum concentration above 0.19 µg/L had a sensitivity of 0.94 and a specificity of 0.70 for predicting mortality after 7 days.

James *et al.* reported S100B serum levels within the first 24 hours after symptom onset to be predictive of neurologic function at hospital discharge in patients with ICH.[44] In a separate study, Delgado *et al.* compared serum levels of S100B in 139 patients with ICH and 776 patients with ischemic stroke within 24 hours of symptom onset. Within the first 6 hours, protein S100B and the receptor for advanced glycation end products (RAGE) best differentiated ICH from ischemic stroke.[45]

Interestingly, an animal study based on a model of collagenase-induced ICH confirmed the rapid increase of S100B in serum in evolving intracerebral hematoma.[46] Peak concentrations were found 6 hours after ICH induction. S100B levels did not decline to values similar to those obtained from sham treated controls until 48 hours had elapsed. All of these studies suggest a rapid release of S100B into the bloodstream in the case of ICH formation.

10.4.1.2 Protein S100B in the Blood of Patients with Ischemic Stroke

An even larger body of evidence suggests release of S100B is delayed in ischemic stroke. Following acute thromboembolic occlusion of arterial brain supply, S100B serum values remain low within the first few hours.[47–49] This correlates well with the histopathological finding that the structural integrity of astroglial cells is well preserved in this early stage of cerebral ischemia.[29–32] In consequence, S100B may not function as a diagnostic tool for acute cerebral ischemia. In the BRAIN study, for example, S100B was not able to differentiate between ischemic stroke and stroke mimics within the first 24 hours of symptom onset.[50] In an analysis based on the NINDS tPA stroke study, no difference was found between S100B levels at hospital admission and at 2 hours after symptom onset. However, at 24 hours, S100B concentrations were modestly increased.[48] The dynamics of S100B increase within 24 hours of stroke onset is highest in cases of large space-occupying ("malignant") infarcts,[51] and serum S100B can be used to predict this devastating type of ischemic stroke.[52] Peak S100B concentrations as well as total S100B release (assessed by means of the area under the curve) are closely correlated with infarct volume and functional outcome.[47,49] Two other studies provide interesting insights into the pathophysiology of S100B release into serum in ischemic stroke.[53,54] Patients with a proximal occlusion of a brain supplying artery (e.g. middle cerebral artery) do not show any S100B release if vessel recanalization occurs early enough in order to prevent the formation of a definite infarction. In other words, a "perfusion deficit" alone, characterized by astroglial cells that temporarily lose their functional but not their structural integrity, does not cause a significant S100B release. S100B is released only when cells become necrotic.

10.4.2 Glial Fibrillary Acidic Protein (GFAP)

GFAP is an astrocytic intermediate filament protein that has important functions in cell migration, process formation, and maintaining cytoskeletal integrity.[55] Similarly to protein S100B, GFAP is abundantly present in the brain. In contrast to protein S100B, however, GFAP is much more brain specific, and thus may have an advantage as a blood biomarker for brain damage and ICH. Extracerebral sources of GFAP are largely unknown. In healthy individuals, GFAP serum concentrations are known to be very low, typically below the lower detection limits of the applied tests.[56,57] Another important difference between GFAP and S100B is that GFAP is not actively released from astroglial cells. Thus, elevated GFAP concentrations in the cerebrospinal fluid or blood may reliably indicate acute BBB damage and astroglial structural disintegration.

10.4.2.1 GFAP in the Blood of Patients with ICH

Meanwhile, considerable data are providing insights in the release kinetics of GFAP after ICH and ischemic stroke. Similar to what is known from S100B measurements, GFAP blood levels increase rapidly in the presence of an expanding ICH, thereby reflecting the immediate destruction of astroglial cells and the BBB. GFAP serum concentrations and ICH volume are closely correlated.[56,58] In one of the first clinical studies, GFAP was measured in patients with acute stroke within 6 hours of symptom onset. GFAP was detected in serum in 81% of patients with ICH, but only in 5% of those with ischemic stroke. Serum GFAP was significantly elevated in patients with ICH compared to patients with ischemic stroke. An optimized cut-off point provided a sensitivity of 79% and a specificity of 98% for differentiating ICH from ischemic stroke.[58] A subsequent study investigated release kinetics of GFAP in patients with ischemic stroke and ICH in more detail to characterize the diagnostic window.[59] Significantly elevated GFAP levels were found as early as 2 to 6 hours after symptom onset in patients with ICH. In contrast, in ischemic stroke, GFAP values remained below the detection limit for the first 12 to 24 hours after symptom onset.

Unden *et al.* found elevated GFAP values in patients with ICH compared to those with ischemic stroke within the first 24 hours after symptom onset.[60] Based on these encouraging findings, a prospective multicenter trial was initiated in order to characterize further the diagnostic potential of GFAP for differentiating ICH and ischemic stroke, respectively.[56] GFAP was measured with a newly developed prototype electro chemiluminometric immunoassay. A total of 205 patients with moderate to severe stroke syndromes (i.e. hemiparesis and signs of hemispheric involvement) were included. Based on brain imaging and clinical evaluation, 163 patients were classified as having ischemic stroke; 39, ICH; and 3, stroke mimics. GFAP concentrations were substantially higher in patients with ICH compared to patients with ischemic stroke (median [interquartile range] 1.91 µg/L [0.41–17.66] versus 0.08 [0.02–0.14], $p < 0.001$).

The overall diagnostic accuracy of GFAP for differentiating ICH from ischemic stroke and stroke mimics was high (AUC = 0.915 [95% CI 0.847–0.982], $p < 0.001$). In contrast to a previous study,[59] but in concordance with the pathophysiological hypothesis,[29] diagnostic accuracy of the GFAP test was found high even within the first 60 minutes after symptom onset (AUC = 0.904). Accuracy of the test was slightly higher in patients who had a severe neurological deficit compared to patients with a moderate stroke syndrome. This is most likely due to the strong relationship between ICH volume and GFAP release, meaning that larger bleeds cause a more pronounced GFAP fingerprint in the blood. The results of this study confirmed the potential of GFAP to function as an early stroke biomarker, allowing rapid discrimination between ICH and ischemic stroke in the emergency setting.

10.4.2.2 GFAP in the Blood of Patients with Ischemic Stroke

Only a few studies have investigated the release kinetics of GFAP into serum in ischemic stroke. In contrast to rapid GFAP release in ICH, GFAP increases in serum are delayed, peaking at day 2 to day 4 after symptom onset. Thus, release kinetics of GFAP are well comparable to that of S100B. In the early phase of ischemic stroke, i.e. within the first 6 hours of symptom onset, GFAP was not detectable in serum (similar to healthy controls).[56,58–60] 12 hours to 24 hours after stroke onset, however, GFAP levels increased steadily and peaked between 48 and 96 hours.[61,62] The amount of GFAP release (assessed by analyzing the area under the curve) is correlated with final infarct size. GFAP levels, therefore, may predict functional outcome and neuropsychological deficits after ischemic stroke.[62] Similar to S100B, kinetics of GFAP release depend on formation of a definite infarction.[63] In other words, a transient cerebral ischemia without tissue necrosis will not lead to detectable GFAP values in serum.

10.5 Discussion

Instantaneous destruction and disintegration of astroglial cells and the BBB observed in the presence of acutely expanding arterial bleeding into the brain parenchyma results in rapid release of intracellular proteins into the extracellular space and the systemic circulation. Serum concentrations of S100B and GFAP, two proteins that are abundantly present in astroglial cells, peak very early in the time course of ICH. In contrast, in ischemic stroke, astroglial necrosis and structural disintegration occur at a later time point, 6 to 12 hours after symptom onset. Thus, within an early time window of stroke, high S100B and GFAP signals in the blood indicate ICH. As GFAP is more brain specific than S100B overall, it may have advantages as a biomarker of acute ICH over S100B and may have a higher diagnostic accuracy for differentiating ICH and ischemic stroke compared to S100B. Smaller amounts of S100B released from the brain in acute parenchymal bleeding may be masked by extracerebral S100B detectable in the bloodstream. In contrast, GFAP has not been found in the serum of healthy individuals, and release of GFAP from the brain into the bloodstream might be more easily detectable compared with protein S100B.

Future availability of a GFAP point-of-care test for rapid identification of patients with ICH would offer crucial advantages over current state-of-the-art practice. This may include optimized triage of patients with acute ICH by emergency physicians towards facilities with intensive care units and neurosurgery departments, whereas patients with ischemic stroke may benefit from rapid admission to a stroke unit.[64,65] Moreover, if current investigations on rapid lowering of blood pressure in patients with acute ICH reveal a benefit not only in terms of ICH expansion but in functional outcome, such therapeutic measures could be applied as early as at the patient's home, after a diagnosis of ICH is established by a positive GFAP test.[66] Furthermore, in patients who are on warfarin anticoagulation while developing symptoms of acute stroke, a positive GFAP test may allow rapid diagnosis of ICH formation. Thus, immediate reversal of anticoagulation by means of fresh frozen plasma or concentrated coagulation factors could be performed to prevent extensive hematoma expansion.[18,19] Despite negative results in a phase III trial, it remains possible that coagulation activation by means of recombinant factor VII reduces ICH volume and improves functional outcome if applied more rapidly after symptom onset.[24] However, as no biomarker-based diagnostic measure can reach a specificity of 100%, which is mandatory to exclude ICH in patients with ischemic stroke subjected to thrombolytic therapy, a GFAP test alone may not suffice in deciding whether to perform thrombolysis in these patients without additional brain imaging.

In summary, astroglial proteins indicate acute ICH in the early phase of stroke. GFAP may open the gate to rapid preclinical differentiation between the two main stroke subtypes and may, for the first time, make it possible to initiate diagnoses-specific therapeutic measures as early as in the paramedic setting.

References

1. N. Badjatia and J. Rosand, *Neurologist*, 2005, **11**, 311.
2. L. Elijovich, P. V. Patel and J. C. Hemphill, *Semin. Neurol.*, 2008, **28**, 657.
3. M. L. Flaherty, D. Woo, M. Haverbusch, P. Sekar, J. Khoury, L. Sauerbeck, C. J. Moomaw, A. Schneider, B. Kissela, D. Kleindorfer and J. P. Broderick, *Stroke*, 2005, **36**, 934.
4. A. I. Qureshi, M. A. Suri, K. Safdar, J. R. Ottenlips, R. S. Janssen and M. R. Frankel, *Stroke*, 1997, **28**, 961.
5. A. I. Qureshi, S. Tuhrim, J. P. Broderick, H. H. Batjer, H. Hondo and D. F. Hanley, *N. Engl. J. Med.*, 2001, **344**, 1450.
6. M. I. Aguilar and W. D. Freeman, *Semin. Neurol.*, 2010, **30**, 555.
7. S. E. Vermeer, A. Algra, C. L. Franke, P. J. Koudstaal and G. J. Rinkel, *Neurology*, 2002, **59**, 205.
8. R. N. Auer and G. R. Sutherland, *Can. J. Neurol. Sci.*, 2005, **32**(Suppl 2), S3.
9. C. M. Fisher, *J. Neuropathol. Exp. Neurol.*, 1971, **30**, 536.

10. S. Takebayashi and M. Kaneko, *Stroke*, 1983, **14**, 28.
11. C. M. Fisher, *Am. J. Pathol.*, 1972, **66**, 313.
12. A. Ellis, *Proc. Pathol. Soc. (Phila.)*, 1909, **12**.
13. C. Kase, J. Mohr and L. Caplan, in *Stroke: Pathophysiology, Diagnosis, and Management*, J. P. Mohr, D. Choi, J. C. Grotta, B. Weir and A. Wolf, New York, Churchill Livingstone, 2004, 4th edition, p. 197.
14. T. I. Mandybur, *J. Neuropathol. Exp. Neurol.*, 1986, **45**, 79.
15. J. P. Vonsattel, R. H. Myers, E. T. Hedley-Whyte, A. H. Ropper, E. D. Bird and E. P. Richardson, *Ann. Neurol.*, 1991, **30**, 637.
16. T. Brott, J. Broderick, R. Kothari, W. Barsan, T. Tomsick, L. Sauerbeck, J. Spilker, J. Duldner and J. Khoury, *Stroke*, 1997, **28**, 1.
17. S. Kazui, H. Naritomi, H. Yamamoto, T. Sawada and T. Yamaguchi, *Stroke*, 1996, **27**, 1783.
18. M. I. Aguilar, R. G. Hart, C. S. Kase, W. D. Freeman, B. J. Hoeben, R. C. Garcia, J. E. Ansell, S. A. Mayer, B. Norrving, J. Rosand, T. Steiner, E. F. Wijdicks, T. Yamaguchi and M. Yasaka, *Mayo Clin. Proc.*, 2007, **82**, 82.
19. T. Steiner, J. Rosand and M. Diringer, *Stroke*, 2006, **37**, 256.
20. C. Foerch, K. Arai, G. Jin, K. P. Park, S. Pallast, K. van Leyen and E. H. Lo, *Stroke*, 2008, **39**, 3397.
21. C. Foerch, K. Arai, E. M. Van Cott, K. van Leyen and E. H. Lo, *J. Cereb. Blood Flow Metab.*, 2009, **29**, 1015.
22. A. Lauer, F. Schlunk, E. M. Van Cott, H. Steinmetz, E. H. Lo and C. Foerch, *J. Cereb. Blood Flow Metab.*, 2010, **31**, 1736.
23. B. B. Thompson, Y. Bejot, V. Caso, J. Castillo, H. Christensen, M. L. Flaherty, C. Foerch, K. Ghandehari, M. Giroud, S. M. Greenberg, H. Hallevi, J. C. Hemphill, 3rd, P. Heuschmann, S. Juvela, K. Kimura, P. K. Myint, Y. Nagakane, H. Naritomi, S. Passero, M. R. Rodriguez-Yanez, J. Roquer, J. Rosand, N. S. Rost, P. Saloheimo, V. Salomaa, J. Sivenius, T. Sorimachi, M. Togha, K. Toyoda, W. Turaj, K. N. Vemmos, C. D. Wolfe, D. Woo and E. E. Smith, *Neurology*, 2010, **75**, 1333.
24. S. A. Mayer, N. C. Brun, K. Begtrup, J. Broderick, S. Davis, M. N. Diringer, B. E. Skolnickand and T. Steiner, *N. Engl. J. Med.*, 2008, **358**, 2127.
25. S. Kazui, K. Minematsu, H. Yamamoto, T. Sawada and T. Yamaguchi, *Stroke*, 1997, **28**, 2370.
26. H. Arima, C. S. Anderson, J. G. Wang, Y. Huang, E. Heeley, B. Neal, M. Woodward, C. Skulina, M. W. Parsons, B. Peng, Q. L. Tao, Y. C. Li, J. D. Jiang, L. W. Tai, J. L. Zhang, E. Xu, Y. Cheng, L. B. Morgenstern and J. Chalmers, *Hypertension*, 2010, **56**, 852.
27. N. Mutlu, R. G. Berry and B. J. Alpers, *Arch. Neurol.*, 1963, **8**, 644.
28. A. I. Qureshi, G. S. Ling, J. Khan, M. F. Suri, L. Miskolczi, L. R. Guterman and L. N. Hopkins, *Crit. Care Med.*, 2001, **29**, 152.
29. R. Brunkhorst, W. Pfeilschifter and C. Foerch, *Transl. Stroke Res.*, 2010, **1**, 246.
30. H. Chen, M. Chopp, L. Schultz, G. Bodzin and J. H. Garcia, *J. Neurol. Sci.*, 1993, **118**, 109.
31. J. H. Garcia, K. F. Liu and K. L. Ho, *Stroke*, 1995, **26**, 636.

32. L. Persson, H. G. Hardemark, H. G. Bolander, L. Hillered and Y. Olsson, *Stroke*, 1989, **20**, 641.

33. J. Aronowski and X. Zhao, *Stroke*, 2011, **42**, 1781.

34. G. Xi, R. F. Keep and J. T. Hoff, *Lancet Neurol.*, 2006, **5**, 53.

35. C. P. Kellner and E. S. Connolly, *Stroke*, 2010, **41**, S99.

36. R. Donato, *Int. J. Biochem. Cell Biol.*, 2001, **33**, 637.

37. R. Donato, *Microsc. Res. Tech.*, 2003, **60**, 540.

38. S. A. Snyder-Ramos, T. Gruhlke, H. Bauer, M. Bauer, A. P. Luntz, J. Motsch, E. Martin, C. F. Vahl, U. Missler, M. Wiesmann and B.W. Bottiger, *Anaesthesia*, 2004, **59**, 344.

39. J. Unden, B. Christensson, J. Bellner, C. Alling and B. Romner, *Scand. J. Infect. Dis.*, 2004, **36**, 10.

40. M. Wiesmann, U. Missler, D. Gottmann and S. Gehring, *Clin. Chem.*, 1998, **44**, 1056.

41. A. Weglewski, D. Ryglewicz, A. Mular and J. Jurynczyk, *Neurol. Neurochir. Pol.*, 2005, **39**, 310.

42. P. Delgado, J. Alvarez Sabin, E. Santamarina, C. A. Molina, M. Quintana, A. Rosell and J. Montaner, *Stroke*, 2006, **37**, 2837.

43. Y. Y. Hu, X. Q. Dong, W. H. Yu and Z. Y. Zhang, *Shock*, 2010, **33**, 134.

44. M. L. James, R. Blessing, B. G. Phillips-Bute, E. Bennett and D. T. Laskowitz, *Biomarkers*, 2009, **14**, 388.

45. P. Delgado, J. Alvarez-Sabin, M. Ribó, F. Purroy, A. Rossell, A. Penalba, I. Fernández-Cadenas, J. Arenillas, C. Molina and J. Montaner, *Oral Presentation, 2005 European Stroke Conference, Bologna, Italy*, 2005.

46. Y. Tanaka, T. Marumo, T. Shibuta, T. Omura and S. Yoshida, *Brain Res. Bull.*, 2009, **78**, 158.

47. C. Foerch, O. C. Singer, T. Neumann-Haefelin, R. du Mesnil de Rochemont, H. Steinmetz and M. Sitzer, *Arch. Neurol.*, 2005, **62**, 1130.

48. E. C. Jauch, C. Lindsell, J. Broderick, S. C. Fagan, B. C. Tilley and S. R. Levine, *Stroke*, 2006, **37**, 2508.

49. U. Missler, M. Wiesmann, C. Friedrich and M. Kaps, *Stroke*, 1997, **28**, 1956.

50. D. T. Laskowitz, S. E. Kasner, J. Saver, K. S. Remmel and E. C. Jauch, *Stroke*, 2009, **40**, 77.

51. W. Hacke, S. Schwab, M. Horn, M. Spranger, M. De Georgia and R. von Kummer, *Arch. Neurol.*, 1996, **53**, 309.

52. C. Foerch, B. Otto, O. C. Singer, T. Neumann-Haefelin, B. Yan, J. Berkefeld, H. Steinmetz and M. Sitzer, *Stroke*, 2004, **35**, 2160.

53. C. Foerch, R. du Mesnil de Rochemont, O. Singer, T. Neumann-Haefelin, M. Buchkremer, F. E. Zanella, H. Steinmetz and M. Sitzer, *J. Neurol. Neurosurg. Psychiatry*, 2003, **74**, 322.

54. M. T. Wunderlich, C. W. Wallesch and M. Goertler, *J. Neurol. Sci.*, 2004, **227**, 49.

55. L. F. Eng, R. S. Ghirnikar and Y. L. Lee, *Neurochem. Res.*, 2000, **25**, 1439.

56. C. Foerch, M. Niessner, T. Back, M. Bauerle, G. M. De Marchis, A. Ferbert, H. Grehl, G. F. Hamann, A. Jacobs, A. Kastrup, S. Klimpe, F.

Palm, G. Thomalla, H. Worthmann and M. Sitzer for the BE FAST Study Group. *Clin. Chem.*, 2012, **58**, 237.

57. U. Missler, M. Wiesmann, G. Wittmann, O. Magerkurth and H. Hagenstrom, *Clin. Chem.*, 1999, **45**, 138.
58. C. Foerch, I. Curdt, B. Yan, F. Dvorak, M. Hermans, J. Berkefeld, A. Raabe, T. Neumann-Haefelin, H. Steinmetz and M. Sitzer, *J. Neurol. Neurosurg. Psychiatry*, 2006, **77**, 181.
59. F. Dvorak, I. Haberer, M. Sitzer and C. Foerch, *Cerebrovasc. Dis.*, 2009, **27**, 37.
60. J. Unden, K. Strandberg, J. Malm, E. Campbell, L. Rosengren, J. Stenflo, B. Norrving, B. Romner, A. Lindgren and G. Andsberg, *J. Neurol.*, 2009, **256**, 72.
61. C. Foerch, O. Singer, T. Neumann-Haefelin, A. Raabe and M. Sitzer, *Cerebrovasc. Dis.*, 2003, **16**(suppl. 4), 45.
62. M. Herrmann, P. Vos, M. T. Wunderlich, C. H. de Bruijn and K. J. Lamers, *Stroke*, 2000, **31**, 2670.
63. M. T. Wunderlich, C. W. Wallesch and M. Goertler, *Eur. J. Neurol.*, 2006, **13**, 1118.
64. D. L. Morris, W. Rosamond, K. Madden, C. Schultz and S. Hamilton, *Stroke*, 2000, **31**, 2585.
65. I. Mosley, M. Nicol, G. Donnan, I. Patrick, F. Kerr and H. Dewey, *Stroke*, 2007, **38**, 2765.
66. C. Delcourt, Y. Huang, J. Wang, E. Heeley, R. Lindley, C. Stapf, C. Tzourio, H. Arima, M. Parsons, J. Sun, B. Neal, J. Chalmers and C. Anderson, *Int. J. Stroke*, 2010, **5**, 110.

CHAPTER 11

Protein S100B in Traumatic Brain Injury

RAMONA ÅSTRAND,*[a] JOHAN UNDÉN[b] AND BERTIL ROMNER[a]

[a] Department of Neurosurgery, Rigshospitalet, Copenhagen, Denmark;
[b] Department of Anesthesia and Intensive Care, Skåne University Hospital, Malmö, Sweden
*Email: raastrand@gmail.com

11.1 Introduction

In 1983, Bakay and Ward suggested that an ideal serum marker should have high specificity for the brain, high sensitivity for brain injury, be released only after irreversible destruction of brain tissue, have rapid appearance in serum, and be released in a time-locked sequence with the injury. They also stated the marker should have a low variability regarding age and gender, there should be reliable and accessible assays for analysis, and there should be clinical relevance.[1] None of the present biomarkers have so far been accepted as an ideal marker for brain injury.

During the past 15–20 years, research around brain biomarkers has accelerated. Most of the work has implicated the protein S100B as a promising surrogate marker for the brain. Various biochemical markers in both cerebrospinal fluid (CSF) and serum have been investigated and research on others is on going.

Protein S100B is the neuromarker that has been the most extensively investigated and has shown the most promising results in head injury management. It is a small (21 kDa) calcium-binding protein expressed mainly

RSC Drug Discovery Series No. 24
Biomarkers for Traumatic Brain Injury
Edited by Svetlana A. Dambinova, Ronald L. Hayes and Kevin K. W. Wang
© Royal Society of Chemistry 2012
Published by the Royal Society of Chemistry, www.rsc.org

in astroglial cells and Schwann cells in the central nervous system. It exerts both intracellular and extracellular effects and, depending on the concentration, it can be either neurotrophic or neurotoxic.[2] Protein S100B can be detected in both CSF and blood. The biomarker concentration has been shown to increase in CSF and/or serum after a vast number of cerebral diseases, e.g. traumatic brain injury (TBI),[3] cerebral infarction,[4] and subarachnoid hemorrhage.[5] Its median concentration in blood in healthy adults has been found to be 0.005 μg/L.[6] The half-life is estimated to be between 90–120 minutes, and it is excreted through the kidneys.[7]

11.2 Protein S100B: Methods of Analysis

During the past few years, various commercially available methods from several manufacturers have been used in the research setting for measuring S100B concentrations in both CSF and serum. Differences between analytical methods should be considered when interpreting results from different studies.[8–10] Commercially available methods include the two-site immunoradiometric assay Sangtec® 100 IRMA; the immunoluminometric assay LIA-mat® Sangtec 100 and LIAISON® Sangtec 100 (DiaSorin AB, Bromma, Sweden); the ELISA Nexus DX™ S100 (SynX Pharma Inc., York, UK); the enzyme-linked immunosorbent assay (Nanogen); the enzyme-linked immunoassay Elecsys® S100; and the more recent electrochemiluminescence immunoassay, Elecsys170 (Roche Diagnostics, Mannheim, Germany).[11,12]

11.3 Protein S100B in Minor Head Injury

11.3.1 Definition of Minor Head Injury

Traumatic brain injury is a significant cause of mortality and morbidity in adults and a leading cause of death in childhood. The diagnostic process includes clinical examination and, in more severe cases, neuroimaging, such as computed tomography of the head (CT) or magnetic resonance imaging (MRI). The majority of patients sustain a minor head injury (>90%), whereas only 5% to 10% are moderate to severe head injuries. The majority of those with minor head injuries, the so-called "minimally head injured" are, according to the Scandinavian Neurotrauma Committee classification system, fully awake and without any loss of consciousness or amnesia post injury and have a very low risk of developing intracranial complications.[13] Those defined as being "mild head injured" are ones who appear fully or nearly fully awake but have had or might have had a brief loss of consciousness and/or amnesia. The mild head injured have an approximately 1% risk of developing intracranial complications in need of neurosurgical intervention; about 5% to 10% have pathological intracranial findings on CT.[14]

11.3.2 Specificity and Sensitivity

During the past decade, several studies have investigated the utility of S100B to identify different types of intracranial pathology and severity. Protein S100B is not specific to the brain; it is also found in low concentrations in adipose tissue, melanocytes in the skin, bone marrow, and heart muscle.[15] Serum S100B increases almost immediately after a relevant brain injury, due to a brief breach of the blood-brain barrier (BBB).[16] Epidural hematoma may initially not be associated with an actual brain damage; therefore, the biomarker might not be increased.[17] However, epidural hematomas reported in the literature have all shown elevated S100B > 0.10 µg/L; albeit only slightly elevated. Thus, S100B could be a marker of BBB disruption as well as brain cell damage.[18]

The sensitivity of the protein is high, making S100B a better marker for identifying patients with a negligible risk for intracranial complications.[19–23] In a recent meta-analysis by Undén *et al.*, the sensitivity for S100B (cut-off 0.10 µg/L and sample time within 3 hours after trauma) was found to be 98% for all types of intracranial pathology on CT. For clinically relevant findings/ neurosurgical intervention, the sensitivity was 100%.[24] It was also estimated that S100B can theoretically reduce the frequency of head CT scanning by 30%, and still maintain patient safety. S100B is not influenced by blood-alcohol levels.[22,25]

11.4 Protein S100B in Severe Traumatic Brain Injury

Research on the potential use of neuromarkers in severe head injury has primarily been directed at severity and outcome prediction. The clinical presentation of an unconscious patient, who is often sedated and/or with external mechanical ventilation, makes traditional clinical evaluation difficult. Initial level of consciousness (Glasgow Coma Scale score), CT classifications (Marshall), pupil response, and events of intracranial hypertension or hypoperfusion/ hypoxia are some of the traditional clinical tools for estimating clinical progress and outcome. The prediction of secondary complications in a neurointensive care patient is also an interesting application of a potential brain biomarker.

11.4.1 Outcome Measures

Several studies have found a correlation between serum S100B levels and clinical outcome according to Glasgow Outcome Scale (GOS) after severe head injury.[26,27] Patients with poor outcome (vegetative state or death) had significantly higher levels of S100B than those with favorable or severely disabled outcomes.[28,29] In a study of 79 patients with severe TBI, a 2.1-fold increase in serum S100B was found in those with unfavorable versus favorable outcomes,[30] and the authors suggested an admission S100B cut-off at 1.13 µg/L for prediction of unfavorable outcome at 6 months. The sensitivity for this was 0.88 and specificity 0.43.

In earlier studies, Raabe *et al.* found a strong association between a cut-off level of 2.5 µg/L (any time) and unfavorable outcome,[26,31] and Woertgen *et al.*

found that serum levels above $2\,\mu g/L$ (drawn within 6 hours after trauma) predicted unfavorable outcome (positive predictive value 87%; negative predictive value, 77%).[32] Nylén *et al.* reported a cut-off of $0.55\,\mu g/L$ (at admission) with a 100% specificity for predicting unfavorable outcome 1 year post injury.[33]

11.4.2 Repeated Measurements

In most studies, S100B sampling is performed at admission and within 24 hours of injury. Some studies have attempted to correlate daily measurements with outcome. Nylén *et al.* correlated maximal serum S100B levels to outcome and found a significant difference in S100B concentrations between patients with favorable and unfavorable outcomes.[33] Böhmer *et al.* collected daily CSF samples in 20 patients with severe TBI and found that early elevations of S100B (up to 3 days) predicted deterioration to brain death.[34] Dimopoulou *et al.* has reported elevated median serum S100B levels of $2.32\,\mu g/L$ in those progressing to brain death, while survivors had significantly lower median values, $1.04\,\mu g/L$.[35] A few studies have investigated prediction of secondary insults after severe TBI, and found a secondary S100B elevation more than 24 hours before clinical deterioration.[36,37] Petzold *et al.* reported that initial serum S100B was able to predict mortality 3 to 4 days before intracranial pressure readings.[38] In a prospective, double-blind, randomized study, Olivecrona *et al.* investigated the prognostic value of biomarkers and concluded that serum is a poor predictor of secondary insults.[39]

Contrary to previous studies, some of the more recent studies have shown no significant correlation between early CSF-S100B or serum S100B and GOS or other outcome scales,[40] nor between dichotomized GOS.[39,41] They conclude there is no clinically significant value of the marker as predictor of clinical outcome.[39]

11.4.3 Multi-Trauma

One drawback of S100B is the presence of extracranial sources, especially bone marrow and adipose tissues.[42] Pelinka *et al.* found that all patients studied with multiple organ trauma demonstrated raised S100B levels whether or not TBI was present, and they concluded that serum levels drawn during the first 24 hours did not reliably predict clinical outcome in these patients. Daily measurements were advocated. However, another study by da Rocha did not find any correlation between higher S100B values and the presence of multi-trauma.[28]

11.5 Protein S100B in Children

There are, by far, fewer studies performed on pediatric TBI and biomarkers, but similar to studies on adults, most studies have focused on the ability of

S100B to predict outcome after TBI. Bechtel and colleagues described use of serum S100B as a screening tool to detect intracranial injuries in children admitted to the emergency department after closed head trauma. Although children with intracranial injuries had significantly higher S100B concentrations than children without intracranial injuries, the ability of S100B levels to detect such injuries was found to be poor.[43]

Child abuse and inflicted TBI is both serious and non-negligible in this patient group. The presenting symptoms are diffuse, such as headache, abdominal pain, and excessive vomiting, and can be difficult to differentiate between other medical conditions, especially among infants. Neither CSF nor serum biomarkers have shown sufficient discriminative abilities to be used as a screening tool for inflicted TBI.[44] In one of the more recent and more encouraging studies from Austria, Castellani *et al.* investigated the correlation of S100B to CT results after mild TBI (GCS 13–15). Serum S100B concentrations were significantly higher in patients with abnormal CT (mean 0.64 µg/L) compared with patients with normal CT (mean 0.50 µg/L) and even though specificity was poor (0.42), the sensitivity and negative predictive value was 1.00. The authors concluded that normal S100B levels (<0.16 µg/L, Elecsys) safely rules out CT pathologies in minor head injured children.[45]

Two independent studies have recently been performed to investigate the reference values of S100B in children, and conclude that healthy children have increased S100B levels compared to adults, and especially among infants.[46,47] These studies encourage further research on this topic to pursue optimizing routines and management in pediatric head injury.

11.6   Health Economics Aspects of S100B in Traumatic Brain Injury

Various guidelines liberally recommend performing a CT scan on patients with mild head injury; however, this is a large patient group and the risk of developing complications is low. Newer guidelines, such as the Canadian CT Head Rule (CCHR) or New Orleans Criteria (NOC) use several clinical criteria for choosing CT after minor head trauma to identify patients with intracranial complications safely, and to reduce unnecessary CT use. However, use of these clinical guidelines is local and is not always properly followed. In addition, evaluation of intoxicated patients is extremely difficult, and the patients are either CT scanned, admitted for observation, or both. Introduction of a biomarker or biomarkers for brain injury would, in theory, reduce the use of unnecessary CT by 30%, when used to supplement the existing guidelines.

To date, the only study on the economic impact of S100B in mild TBI has been published by Ruan *et al.*, and concluded that use of S100B will lower costs only if head CT scan rates for patients with isolated mild TBI are relatively high and if blood tests require less time than imaging.[48] The analysis was based

on adult patients with isolated minor head injury with a GCS score of 15. Sensitivity analyses were based on information as reported in the literature, which makes the reliability less strong. Protein S100B has since been introduced into clinical use after mild head injury and the evaluation of its economic impact is still on going.

The health economic aspects of a biomarker in pediatric minor head injury should, in theory, have even greater impact than for adults. The use of head CT is recommended very liberally in children, although the effective dose of radiation is higher for children than for adults. Children also have a higher lifetime risk of developing malignancies due to CT radiation, but so far the use of CT in children is only increasing.[49] A head CT is more cost-effective than admission for observation for 24–48 hours. Thus, any possibility of reducing the use of unnecessary CT would be most welcome.

References

1. L. Bakay, *World J. Surg.*, 1983, **7**, 42.
2. R. Donato, *Microsc. Res. Tech.*, 2003, **60**, 540.
3. T. Ingebrigtsen, K. Waterloo, E. A. Jacobsen, B. Langbakk and B. Romner, *Neurosurgery*, 1999, **45**, 468, discussion 475.
4. M. Herrmann, P. Vos, M. T. Wunderlich, C. H. de Bruijn and K. J. Lamers, *Stroke*, 2000, **31**, 2670.
5. S. Moritz, J. Warnat, S. Bele, B. M. Graf and C. Woertgen, *J. Neurosurg. Anesthesiol.*, 2010, **22**, 21.
6. M. Wiesmann, U. Missler, D. Gottmann and S. Gehring, *Clin. Chem.*, 1998, **44**, 1056.
7. H. Jonsson, P. Johnsson, P. Hoglund, C. Alling and S. Blomquist, *J. Cardiothorac. Vasc. Anesth.*, 2000, **14**, 698.
8. K. Muller, A. Elverland, B. Romner, K. Waterloo, B. Langbakk, J. Unden and T. Ingebrigtsen, *Clin. Chem. Lab. Med.*, 2006, **44**, 1111.
9. M. Hallen, R. Carlhed, M. Karlsson, T. Hallgren and M. Bergenheim, *Clin. Chem. Lab. Med.*, 2008, **46**, 1025.
10. B. Alber, R. Hein, C. Garbe, U. Caroli and P. B. Luppa, *Clin. Chem. Lab. Med.*, 2005, **43**, 557.
11. R. Harpio and R. Einarsson, *Clin. Biochem.*, 2004, **37**, 512.
12. C. W. Heizmann, *Clin. Chem.*, 2004, **50**, 249.
13. T. Ingebrigtsen, B. Romner and C. Kock-Jensen, *J. Trauma*, 2000, **48**, 760.
14. M. Smits, D. W. Dippel, G. G. de Haan, H. M. Dekker, P. E. Vos, D. R. Kool, P. J. Nederkoorn, P. A. Hofman, A. Twijnstra, H. L. Tanghe and M. G. Hunink, *JAMA*, 2005, **294**, 1519.
15. T. O. Kleine, L. Benes and P. Zofel, *Brain Res. Bull.*, 2003, **61**, 265.
16. N. Marchi, P. Rasmussen, M. Kapural, V. Fazio, K. Kight, M. R. Mayberg, A. Kanner, B. Ayumar, B. Albensi, M. Cavaglia and D. Janigro, *Restor. Neurol. Neurosci.*, 2003, **21**, 109.

17. J. Unden, J. Bellner, R. Astrand and B. Romner, *Br. J. Neurosurg.*, 2005, **19**, 43.
18. J. Unden and B. Romner, *Scand. J. Clin. Lab. Invest.*, 2009, **69**, 13.
19. P. Biberthaler, T. Mussack, E. Wiedemann, T. Gilg, M. Soyka, G. Koller, K. J. Pfeifer, U. Linsenmaier, W. Mutschler, C. Gippner-Steppert and M. Jochum, *Shock*, 2001, **16**, 97.
20. P. Biberthaler, T. Mussack, E. Wiedemann, K. G. Kanz, M. Koelsch, C. Gippner-Steppert and M. Jochum, *World J. Surg.*, 2001, **25**, 93.
21. T. Ingebrigtsen, B. Romner, S. Marup-Jensen, M. Dons, C. Lundqvist, J. Bellner, C. Alling and S. E. Borgesen, *Brain Inj.*, 2000, **14**, 1047.
22. T. Mussack, P. Biberthaler, K. G. Kanz, U. Heckl, R. Gruber, U. Linsenmaier, W. Mutschler and M. Jochum, *Shock*, 2002, **18**, 395.
23. L. F. Poli-de-Figueiredo, P. Biberthaler, C. Simao Filho, C. Hauser, W. Mutschler and M. Jochum, *Clinics (Sao Paulo)*, 2006, **61**, 41.
24. J. Unden and B. Romner, *J. Head Trauma Rehabil.*, 2010, **25**, 228.
25. P. Biberthaler, T. Mussack, E. Wiedemann, K. G. Kanz, T. Gilg, C. Gippner-Steppert and M. Jochum, *Acta Neurochir. Suppl.*, 2000, **76**, 177.
26. A. Raabe, C. Grolms and V. Seifert, *Br. J. Neurosurg.*, 1999, **13**, 56.
27. K. Nylen, M. Ost, L. Z. Csajbok, I. Nilsson, K. Blennow, B. Nellgard and L. Rosengren, *J. Neurol. Sci.*, 2006, **240**, 85.
28. A. B. da Rocha, R. F. Schneider, G. R. de Freitas, C. Andre, I. Grivicich, C. Zanoni, A. Fossa, J. T. Gehrke, G. Pereira Jotz, M. Kaufmann, D. Simon and A. Regner, *Clin. Chem. Lab. Med.*, 2006, **44**, 1234.
29. M. Wiesmann, E. Steinmeier, O. Magerkurth, J. Linn, D. Gottmann and U. Missler, *Acta Neurol. Scand.*, 2010, **121**, 178.
30. P. E. Vos, B. Jacobs, T. M. Andriessen, K. J. Lamers, G. F. Borm, T. Beems, M. Edwards, C. F. Rosmalen and J. L. Vissers, *Neurology*, 2010, **75**, 1786.
31. A. Raabe, C. Grolms, O. Sorge, M. Zimmermann and V. Seifert, *Neurosurgery*, 1999, **45**, 477.
32. C. Woertgen, R. D. Rothoerl, C. Metz and A. Brawanski, *J. Trauma*, 1999, **47**, 1126.
33. K. Nylen, M. Ost, L. Z. Csajbok, I. Nilsson, C. Hall, K. Blennow, B. Nellgard and L. Rosengren, *Acta. Neurochir. (Wien)*, 2008, **150**, 221, discussion 227.
34. A. E. Bohmer, J. P. Oses, A. P. Schmidt, C. S. Peron, C. L. Krebs, P. P. Oppitz, T. D'Avila, D. O. Souza, L. V. Portela and M. A. Stefani, *Neurosurgery*, 2011, **68**, 1624.
35. I. Dimopoulou, S. Korfias, U. Dafni, A. Anthi, C. Psachoulia, G. Jullien, D. E. Sakas and C. Roussos, *Neurology*, 2003, **60**, 947.
36. A. Raabe and V. Seifert, *J. Neurosurg.*, 1999, **91**, 875.
37. L. E. Pelinka, E. Toegel, W. Mauritz and H. Redl, *Shock*, 2003, **19**, 195.
38. A. Petzold, A. J. Green, G. Keir, S. Fairley, N. Kitchen, M. Smith and E. J. Thompson, *Crit. Care Med.*, 2002, **30**, 2705.

39. M. Olivecrona, M. Rodling-Wahlstrom, S. Naredi and L. O. Koskinen, *J. Neurol. Neurosurg. Psychiatry*, 2009, **80**, 1241.
40. B. M. Bellander, I. H. Olafsson, P. H. Ghatan, H. P. Bro Skejo, L. O. Hansson, M. Wanecek and M. A. Svensson, *Acta. Neurochir. (Wien)*, 2011, **153**, 90.
41. J. Unden, R. Astrand, K. Waterloo, T. Ingebrigtsen, J. Bellner, P. Reinstrup, G. Andsberg and B. Romner, *Neurocrit Care*, 2007, **6**, 94.
42. J. Unden, J. Bellner, M. Eneroth, C. Alling, T. Ingebrigtsen and B. Romner, *J. Trauma*, 2005, **58**, 59.
43. K. Bechtel, S. Frasure, C. Marshall, J. Dziura and C. Simpson, *Pediatrics*, 2009, **124**, e697.
44. R. P. Berger, P. D. Adelson, M. C. Pierce, T. Dulani, L. D. Cassidy and P. M. Kochanek, *J. Neurosurg.*, 2005, **103**, 61.
45. C. Castellani, P. Bimbashi, E. Ruttenstock, P. Sacherer, T. Stojakovic and A. M. Weinberg, *Acta. Paediatr.*, 2009, **98**, 1607.
46. R. Astrand, B. Romner, J. Lanke and J. Unden, *Clin. Chim. Acta*, 2011, **412**, 2190.
47. C. Castellani, T. Stojakovic, M. Cichocki, H. Scharnagl, W. Erwa, A. Gutmann and A. M. Weinberg, *Clin. Chem. Lab. Med.*, 2008, **46**, 1296.
48. S. Ruan, K. Noyes and J. J. Bazarian, *J. Neurotrauma*, 2009, **26**, 1655.
49. D. Brenner, C. Elliston, E. Hall and W. Berdon, *AJR. Am. J. Roentgenol.*, 2001, **176**, 289.

CHAPTER 12

Utilities of TBI Biomarkers in Various Clinical Settings

STEFANIA MONDELLO,*[a,b] RONALD L. HAYES[a,b] AND KEVIN K. W. WANG[a,b]

[a] Banyan Biomarkers, 13400 Progress Blvd., Alachua, FL 32615, USA;
[b] University of Florida, 1600 SW Archer Road, Box 100254, Gainesville, FL, USA
*Email: smondello@banyanbio.com

12.1 Introduction

Traumatic brain injury (TBI) is a major health and socioeconomic problem that affects all societies. However, traditional approaches to the classification of clinical severity are the subject of debate and are being supplemented with structural and functional neuroimaging, as the need for biomarkers that reflect elements of the pathogenetic process is widely recognized. Basic science research and developments in the field of proteomics have greatly advanced our knowledge of the mechanisms involved in damage and have led to the discovery and rapid detection of new biomarkers that were not available previously. However, translating this research for the benefit of patients remains a challenge. In this article, we summarize new developments, current knowledge, and controversies, focusing on the potential role of these biomarkers as diagnostic, prognostic, and monitoring tools of brain-injured patients.

12.2 Biomarker Discovery: Methods and Results

Novel and powerful technologies hold special promise for the discovery of novel biomarkers that might form the foundation for new clinical blood tests with unprecedented accuracy and sensitivity. We will review proteomics

RSC Drug Discovery Series No. 24
Biomarkers for Traumatic Brain Injury
Edited by Svetlana A. Dambinova, Ronald L. Hayes and Kevin K. W. Wang
© Royal Society of Chemistry 2012
Published by the Royal Society of Chemistry, www.rsc.org

techniques applied for the first time to discovery of biomarkers of the central nervous system (CNS). These techniques include refined mass spectrometry technology and high throughput immunoblot techniques. Results of these approaches can identify potential candidate biomarkers employing systems biology and data-mining methods, which will also be described.

12.2.1 Proteomics/Systems Biology in the Area of Neurotrauma

The application of neuroproteomics/neurogenomics has revolutionized the characterization of protein/gene dynamics, leading to a greater understanding of post-injury biochemistry. The neuroproteomics and neurogenomics fields have undertaken major advances in the area of neurotrauma research focusing on biomarker identification. Several candidate markers have been identified and are being evaluated for their efficacy as biological biomarkers utilizing these "omics approaches." The identification of these differentially expressed candidate markers using these techniques is proving to be only the first step in the biomarker development process. However, to translate these findings into the clinic, a data-driven development cycle incorporating data-mining steps for discovery, qualification, verification, and clinical validation are needed. Data-mining steps extend beyond the collected data level into an integrated scheme of animal modeling, instrumentation, and functional data analysis.

Proteomics is the identification and quantification of all expressed proteins of a cell type, tissue, or organism. The advancement in the field of proteomics has coincided with the completion of the human genome sequencing project.[1] In recent years, the term "proteomics" is often mentioned together with biomarker discovery, as proteomic studies have the capability of identifying sensitive and unique signature protein biomarkers from tissues or biofluids derived from animal models or human clinical samples inflicted with various diseases. Neuroproteomics and neurogenomics, the application of proteomics and genomics in the field of neuronal injury, have been identified as a potential means for biomarker discovery, with the ability to identify proteome dynamics in response to brain injury.[2-6]

In the area of brain injury, several studies have demonstrated the role of proteomics[7,8] and genomics[9,10] in providing significant insight into understanding changes, modifications, and functions in certain proteins post TBI. In addition, genomics and proteomics are powerful, complementary tools that play an important role in the area of biomarker identification. Over the past few years, advances in the fields of neuroproteomics and neurogenomics have led to the discovery of many candidate biomarkers and are becoming the primary methods for initial candidate marker selection.[11-15] The identification of differentially expressed candidate markers using these techniques is proving to be only the first step in the biomarker development process. However, to translate these into the clinic, these novel assays require a data-driven development cycle that incorporates data-mining steps for discovery, qualification, verification, and clinical validation.[16]

TBI neuroproteomics studies have utilized biofluids such as blood/serum in addition to injured tissue to identify clinical markers that may correlate with injury severity. One of the studies, by Burgess *et al.*, evaluated altered differential proteins in normal human postmortem cerebrospinal fluid (CSF).[17] The rationale of using postmortem CSF is that it resembles a model of massive brain injury and cell death; thus, comparing the protein profile of postmortem CSF with brain injury CSF would be ideal for identifying protein markers of injury. Of the 229 proteins identified, a total of 172 were novel and not previously described. The findings showed that the use of postmortem CSF (non-TBI samples) to evaluate altered protein levels mimicked the changes occurring in the brain following a traumatic insult. Furthermore, the identification of differential proteins of intracellular origin in the CSF corroborates the suggestion that there is protein leakage into the CSF following brain injury.[18,19] This is a key step in identifying protein markers, since neuronal-specific proteins leak from the injured brain directly into the CSF.

In one of the TBI studies conducted in our laboratory, 1D-differential gel electrophoresis (DIGE) protein separation in series with mass spectrometry analysis was used to discover putative TBI biomarkers in brain tissues from a rat model.[5] These included 57 down-regulated and 74 up-regulated proteins; however, the data were not so informative due to limited separation capability. In an advanced study, we utilized a multidimensional separation platform called CAX-PAGE/RPLC-MSMS, which comprised different levels of separation, including ion chromatography, 1D gel electrophoresis, and mass spectrometry as a novel approach for identifying biomarkers and protein breakdown products (degradomes); for detailed reviews, refer to Kobeissy *et al.*[20] and Chen *et al.*[21] As an application, the CAX/neuroproteomic analysis was employed on cortical samples of rat subjected to a controlled cortical impact (CCI) model of experimental TBI (48 hours post injury). Of interest, our neuroproteomic analysis identified 59 differential protein components, of which 21 decreased and 38 increased in levels after TBI. One main advantage of this technique is its ability to elucidate degradomic substrates of different protease systems; thus, our data identified the elevated levels of the breakdown products of several proteins.[21] Several of these are now being investigated as potential biomarkers specific for TBI to assess severity and recovery by evaluating their levels at different time points post TBI.

12.2.2   Data Mining Coupled Neurosystems Biology Analysis in Brain Injury

Coupled to data-mining steps, systems biology (SB) represents a mathematical model capable of predicting the altered processes or functions of a complex system under normal and perturbed conditions. It combines experimental, basic science data sets, proteomic and genetic data sets, literature and text mining, integration with computational modeling, bioinformatics, and pathway/interaction mapping methods. When constructed properly, SB databases

can provide a context or framework for understanding biological responses within physiological networks at the organism level, rather than in isolation.[22]

In this regard, "omics" output constitutes one key component of neurosystems biology. It discusses the global changes involved in neurological perturbations, integrating the final outcomes into a global functional network map which incorporates potential biomarkers identified.[23,24] In the area of brain injury, the neurosystems biology platform harnesses data sets that, by themselves, would be overwhelming, into an organized, interlinked database that can be queried to identify non-redundant brain injury pathways or convert hot spots. These can be exploited to determine their utilities as diagnostic biomarkers and/or therapeutic targets. The ultimate goals of system biology are: first, by exploring the systems component (e.g. gene, protein, small molecule, metabolite), biologists, pharmaceutical companies, and clinicians are better able to understand the mechanisms underlying disease components. Thus, it allows for suitable targets for treatment. Secondly, the systems biology approach enables one to be able to predict the functions and behavior of various components of the system upon varying any of the interconnected components, since the whole system will be viewed globally rather than on a micro, individual component level.[25]

In the field of neurotrauma, identifying and analyzing brain injury-related networks provides important and practical clues relating to biological pathways relevant to disease processes. However, the more important underlying goal in this analysis is to provide important details that may suggest radically new approaches to therapeutics. Systems modeling and simulation is now considered fundamental to the future development of effective therapies. In brain injury, for example, it has been shown that calpain and caspase proteases are major components in cell death pathways taking part in two destructive proteolytic pathways that not only contribute to key forms of cell death (necrosis and apoptosis), but also in the destruction of important structural components of the axons (alpha II-spectrin breakdown products [SBDPs] and tau), dendrites (MAP2), and myelin (MBP) (Figure 12.1). Interestingly, two different forms of SBDPs reflect either neuronal necrosis (SBDP150 and SBDP145 cleaved by calpain) or neuronal apoptosis (SBDP120 cleaved by caspase-3).[26] These SBDPs and other similar neural protein breakdown products can serve as target pathway specific biomarkers, as illustrated in Figure 12.1.

12.3 Strategy for Regulatory Approval by FDA

To date, studies of biomarkers for brain injury have been restricted to research applications only. Although there is broad recognition of clinical utility of biomarkers, the U.S. Food and Drug Administration (FDA) has yet to approve any biomarkers of CNS injury or disease. This section provides a detailed outline of the regulatory consideration necessary for a biomarker to file for FDA approval. Importantly, many of these considerations need to be

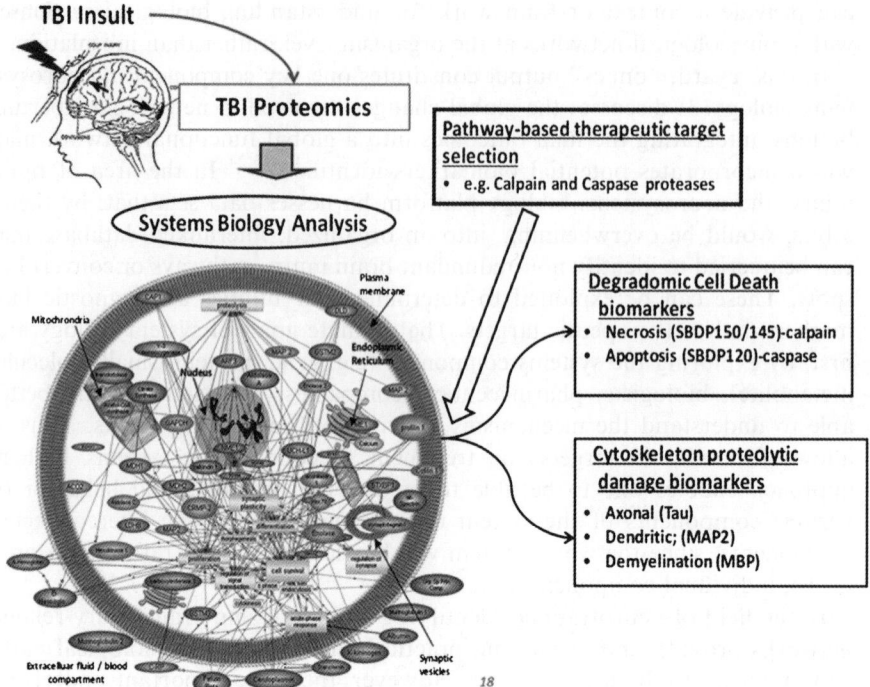

Figure 12.1   Systems biology-based therapeutic target identification and target-specific biomarker selection. Calpain and caspase proteases are used here as examples of therapeutic targets with proteolytic brain biomarkers representing non-redundant pathways relevant to the pathobiology of these therapeutic targets and the disease itself. TBI (traumatic brain injury); MAP2 (microtubule-associated protein 2); MBP (myelin basic protein); SBDP (spectrin breakdown product).

integrated into relatively early stages of biomarker validation, assay development, and device selection.

12.3.1   FDA Regulation

The FDA has been involved in the regulation of medical devices, including *in vitro* diagnostic devices (IVDs) since 1976, when the U.S. Congress established laws for the regulation of medical devices under the Medical Device Amendments act.[27] According to the Medical Device Amendments act, for an IVD to enter the U.S. market, it must comply with a set of rules and regulations in order to prove safety and effectiveness for its intended use. In 21 CFR 860.7(d)(1), device safety requires "the probable benefit to health from use of device for its intended use and conditions of use … outweigh any probable risk". In 21 CFR 860.7(e)(1), device effectiveness requires "that in a significant portion of the target population, the use of the device for its intended use and conditions for use ... will provide clinically significant results". In addition, a new assay is required to demonstrate an adequate analytical performance (appropriate accuracy and precision) and clinical performance (sensitivity, specificity, and some indication of clinical utility) (21 CFR 807; 21 CFR 814).[28,29]

To be commercialized in a kit, newly discovered biomarkers device must follow specific pathways. The first step, which consists of investigational studies of diagnostic devices, can be performed with various design configurations, but they must conform to FDA requirements. Researchers are required to apply for an investigational device exemption (IDE) before initiating the study, as described in Title 21 of the Code of Federal Regulations 812 (21 CFR 812).[30] The IDE submission should describe the nature of the proposed study, include details of informed consent, and ensure patient protection and that the risks associated with participation in the study will be clearly communicated to individuals.

After the appropriate investigational studies have been completed, the FDA requires premarket submissions before a test can be approved for clinical use in the United States. FDA considers three classes of devices (class I, II, or III),[31,32] Depending on the nature of the test and its classification, the product could be reviewed as a "510(k) premarketing clearance" with a 90-day timeline or "premarket approval (PMA)" with a 180-day timeline (21 CFR 807; 21 CFR 814).[28,29]

The 510(k) process is used when the new test measures an existing FDA classified analyte (class I or II) where there exists a commercially available predicate test method that has been cleared by the FDA or that was in commercial distribution before May 28, 1976 (21 CFR 860.84). Premarket clearance requires the sponsor to provide information for the new product, including its intended use and classification and "substantially equivalence" to the predicate device. This ensures that a high level of safety and effectiveness is maintained. In addition, the sponsor must show characterization of analytical capability of the test (e.g. specificity and accuracy, precision, and linearity by correlating patient studies against the predicate device) (http://www.fda.gov/MedicalDevices/DeviceRegulationandGuidance/Overview/MedicalDeviceUserFeeandModernizationActMDUFMA/default.htm).

The PMA process is used when the test is classified as class III; that is, either it is associated with high risk (e.g. when the outcome determines cancer treatment or diagnosis) or the clinical utility of the marker or the technology of the measurement are novel and no predicate device can be identified (21 CFR 814).[29] The FDA requires the sponsor to submit the same data required for 510(k) as well as clinical outcomes data, where the level of the marker is related to disease status defined by clinical criteria (http://www.fda.gov/MedicalDevices/DeviceRegulationandGuidance/Overview/MedicalDeviceUserFeeandModernizationActMDUFMA/default.htm) (Figure 12.2).

In 1997, since some new biomarkers have no obvious predicate devices and do not have safety concerns, FDA created a new hybrid "de novo" or "risk-based" classification (21 CFR 814).[29] This process allows a new biomarker to be regulated as in a 510(k), but requires the demonstration of clinical effectiveness.

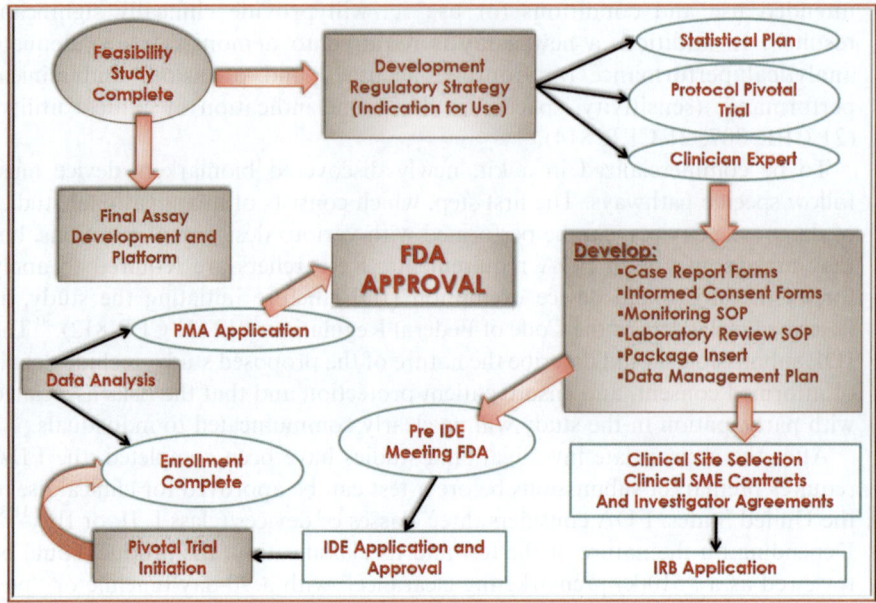

Figure 12.2   Regulatory pathway for PMA application.

Therefore, in the absence of a predicate device and depending on the intended use and clinical utility, either a PMA or a de novo process will be required for a new protein of interest before it achieves commercial availability in a kit or device.

12.3.2   Quality System Regulation Requirements

An under-appreciated challenge to marketing a biomarker assay for clinical use is conforming with the current good manufacturing practices (cGMP) requirements set forth by the FDA for quality system regulation (QS regulation). Medical devices must be designed, manufactured, packaged, labeled, stored, installed, and serviced according to these requirements. Academic researchers or clinical scientists who develop biomarker assays are often not familiar with QS regulation. Although QS regulation is not different in principle from the good laboratory practice that is commonly used in clinical laboratories, it is more regimented and based on the concept of up-front delineation of device design. This requires that, before making a new test, sponsors must identify relevant inputs (features of how a test is performed) and outputs (expected performance requirements), and must establish (define, document, and implement) a program for verifying and validating production processes and test performance. Once a product has been developed without attention to these details, collecting the data necessary for regulatory approval might be impossible or duplicative. Researchers often fail to realize that documentation, quality control, and specific definition and explanation of design features are required by all regulatory bodies. Research institutes or others that are developing new diagnostic assays without experience of meeting regulatory requirements are urged to consider either hiring or contracting to obtain the expertise necessary. The built-in quality from following these regulations is likely to be worth the initial costs, and the end product is likely to be one that is clinically valid and meets user needs in a manner that will help to ensure the acceptance of a new assay.

12.4 Biomarker-based Diagnosis, Management, and Outcome Prediction of Patients with Severe TBI

Traumatic injury to the brain results in cellular activation and disintegration, leading to release of cell type-specific proteins. Measurable amounts of these damage markers are present in the CSF and blood. These markers not only indicate the pathoanatomic injury type and the severity of injury, but also might provide specific information about the pathophysiologic mechanisms that can be targeted by therapeutic interventions. Therefore, during the past decade, neurobiochemical markers for TBI have attracted increased attention because they can be used to screen for, diagnose, or monitor patients and guide targeted therapy or assess therapeutic response.[33,34] Furthermore, biomarkers might be valuable tools in drug development[35] (Figure 12.3).

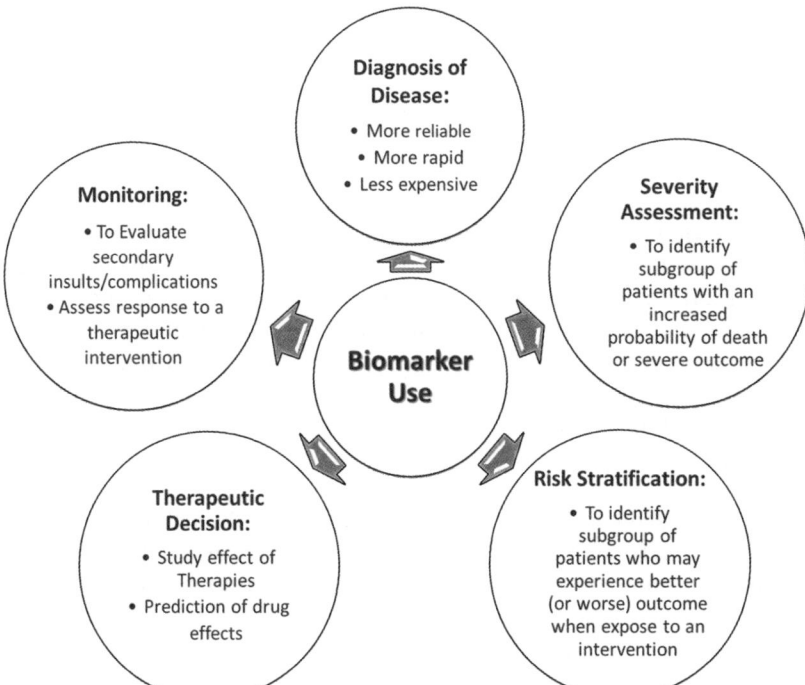

Figure 12.3 Potential use of biomarkers in TBI.

Many publications can be found for candidate biomarkers, although initially promising results have often not been confirmed. Here we review novel CSF and serum biomarkers for TBI. We focus on established biomarkers that have shown high sensitivity and specificity in at least two independent studies and discuss their potential role in the clinic.

12.4.1 Primary Damage and Acute Diagnosis in Severe TBI

External forces, as a consequence of direct impact, rapid acceleration or deceleration, or blast waves from an explosion or a penetrating object (e.g. gunshot) can result in one of several pathologies included in the heterogeneous and complex spectrum defined as brain injury. The unifying factor is that the initial injury results in cellular damage and disintegration, leading to release of cell type-specific proteins such as the calcium-binding protein S100B, myelin basic protein (MBP), glial fibrillary astrocytic protein (GFAP), ubiquitin C-terminal hydrolase (UCH-L1), and neuron-specific enolase (NSE). Based on this characteristic, biomarkers have been used in the acute phase to assess severity of injury, correlation with clinical examination, and neuroradiological findings/types of injury. Several reports suggested that S-100B, a calcium-binding protein primarily expressed in astrocytes, might be valuable in acute phase in TBI.[36,37] Consistently, S-100B serum levels have been reported to correlate with both GCS scores and neuroradiological findings at hospital admission.[36,37,39] However, the lack of specificity[38,40–42] has tempered the original enthusiasm regarding the usefulness of this protein as a brain damage biomarker. Because of the important function and its high brain specificity, UCH-L1, a neuron-specific cytoplasmic enzyme, has been proposed as a novel biomarker for TBI. A recent study systematically assessed UCH-L1 in human serum following TBI for the first time and compared levels with those found in CSF of the same patients.[43] UCH-L1 was detectable in the blood very early after injury. As shown by this work and others,[43,44] the temporal profile of changes in biomarker levels is an important factor in determining diagnostic utility (Figure 12.4). Levels of UCH-L1 have been found to be significantly increased in patients with severe TBI compared with uninjured patients, and a significant association between UCH-L1 concentrations and injury magnitude was also observed.[43,45] From the same cohort, a study investigating the exposure and kinetic metrics of UCH-L1 found a strong correlation between CSF and serum exposure and kinetic characteristics, especially during the acute period.[46]

Relevant to acute diagnosis in TBI, in a report by Mondello *et al.*,[47] different patterns of UCH-L1 and GFAP release in serum were associated with different types of brain-injured patients as characterized by neuroimaging (diffuse injury versus focal mass lesion), including the finding that the two biomarkers had different temporal profiles in the same type of injury (Figure 12.5). This suggests that different biomarker pathways involve different patterns of structural damage and different pathophysiological mechanisms. A key question in this regard is whether the relationship among protein biomarkers could provide

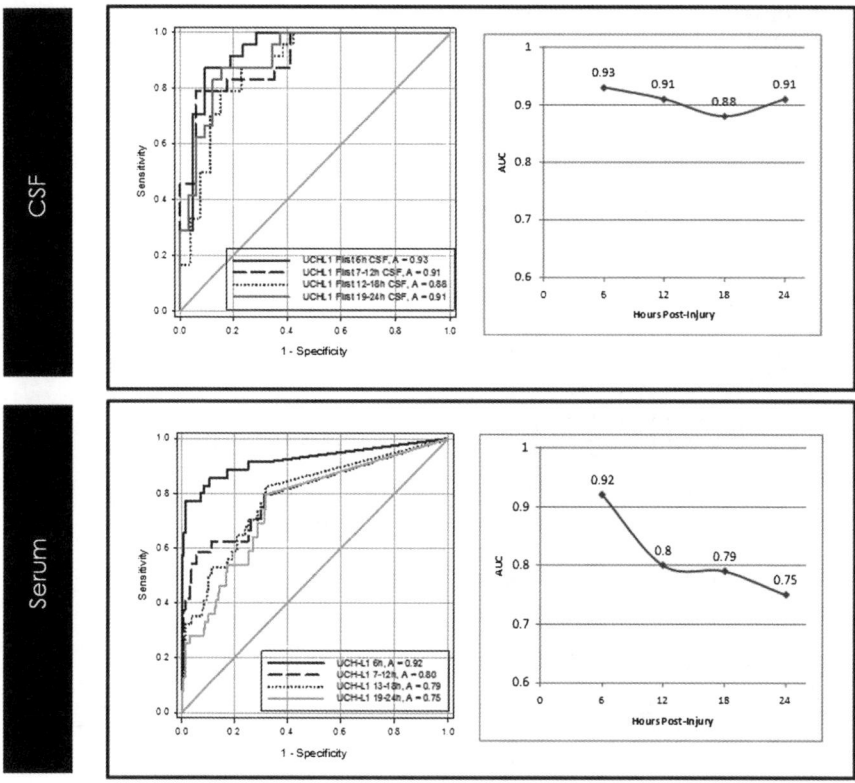

Figure 12.4 UCH-L1 receiver operation characteristics (ROC) curves. Comparing 24 h post injury TBI CSF and serum UCH-L1 level versus CSF controls or serum normal controls, respectively. Area under the curve (AUC) over time was also plotted. Adapted from Mondello *et al.*[42]

better characterization of subjects at risk for specific types of cellular damage who may require different therapeutic approaches.

Recent research has raised new insights in the pathophysiological mechanism of TBI. In 2007, Pineda *et al.*[48] assessed calpain versus caspase cleavage products of α-spectrin in human CSF from subjects with severe TBI, under the premise that the former reflects cell death from necrosis, while the latter reflects cell death from apoptosis. Consistently, works[49,50] assessing serial CSF samples from adults with severe TBI compared calpain versus caspase cleavage products of α-spectrin and also suggested that necrosis may predominate in TBI since, despite marked early increases in calpain cleavage products in CSF, there were only low levels of caspase cleavage products.

A recent addition to the growing number of markers of acute brain injury is the heavily phosphorylated form of the major neurofilament subunit NF-H (pNF-H). Because pNF-H is specific for axons, increased serum concentrations of this protein are expected to provide a specific measure of ongoing axonal damage or degeneration.[51,52]

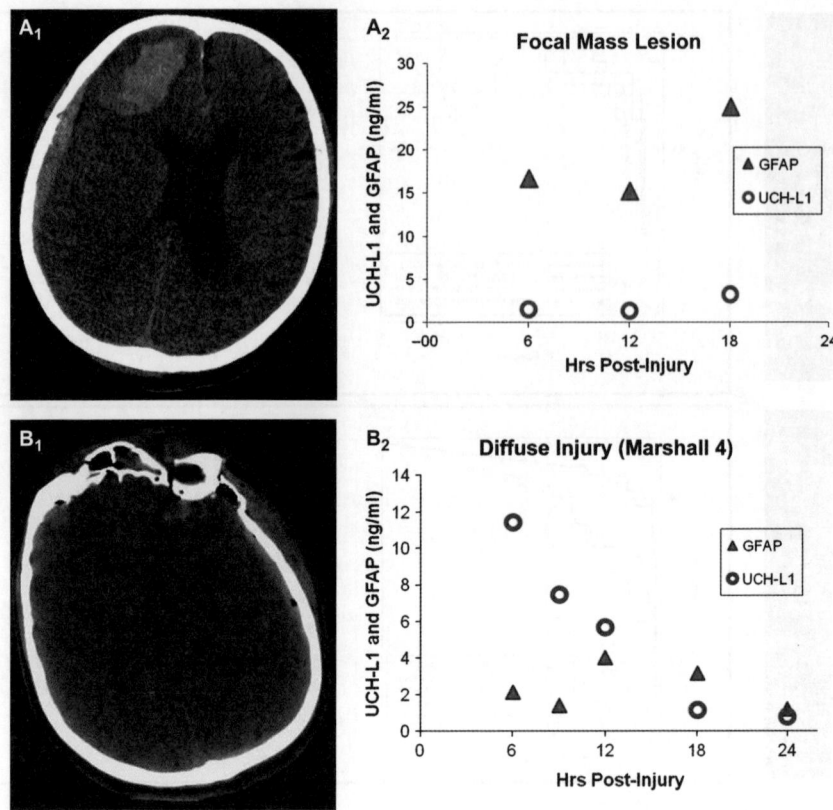

Figure 12.5 Ubiquitin C-terminal hydrolase-L1 protein (UCH-L1) and glial fibrillary acidic protein (GFAP) dynamics in human serum after acute injury. A_1 and B_1: computed tomography scans demonstrating focal and diffuse brain injury in individual patients. C_1 and D_1: UCH-L1 (green circles) and GFAP (blue triangles) serum concentrations measured every 6 hours in corresponding individual patients. Time (*x*-axis) reflects interval after injury. Adapted from Mondello *et al.*[46]

12.4.2   Biomarkers in the Management of Severe TBI

Management of severe TBI patients can importantly influence outcome. A major focus for neurointensive care is to prevent and limit ongoing brain damage and to provide the best conditions for natural brain recovery by reducing the number of secondary insults. Optimum oxygenation, perfusion, nutrition, glycemic control, and temperature homoeostasis are indicated, as in general intensive care. Furthermore, the brain must be protected from overt or silent seizures. Although a number of researchers have focused on identifying biomarkers to diagnose injury severity and survivability, only a few groups have focused on identifying prognostic biomarkers for these secondary pathologies. The discovery of biomarkers for these secondary injuries would greatly aid in the design of management strategies for individual patients. The serum levels of GFAP and S100B have been proposed as markers capable of differentiating between TBI patients based on their ICP status. Raabe *et al.*[36] reported that the serum peak value of S-100B was associated with the lowest cerebral perfusion pressure (CPP) value and the highest intracranial pressure (ICP). Pelinka and colleagues,[53] evaluating serum GFAP and S100B levels in 92 patients with TBI, found that these proteins were significantly elevated in patients with > 25 mmHg. In a prospective study evaluating the relationship of cytokine levels with ICP and CPP in 24 patients with severe TBI, Stein *et al.*[54] reported that serum IL-8 and TNF-α levels could be used to predict impending secondary insults, namely intracranial hypertension (ICH) and cerebral hypoperfusion (CH), before the clinical manifestation of these events. Given the known morbidity of ICH and CH, early intervention and prevention may have a significant impact on outcome.

Furthermore, the clinical introduction of microdialysis brought hope for new insights into the neurochemistry of acute human brain injury and the possibility of neurochemical monitoring in neurointensive care management. The clinical value of biomarkers may increase when microdialysis is used to harvest the biomarkers directly in the injured brain parenchyma, because of an improved spatial and temporal resolution in comparison with CSF and blood analysis. To date, analytes including inflammatory biomarkers such as cytokines and protein biomarkers are investigational, but hold promise for future application in advancing our understanding of basic pathophysiology, patient management, and therapeutic target selection.

12.4.3 Biomarkers for Prediction of Outcome and Prognosis

Experience shows that about 85% of recovery occurs within 6 months after injury, but further recovery can occur later. Early and intensive rehabilitation is recommended to achieve the best possible functional outcome and social reintegration. However, the optimum timing and approach to rehabilitation of patients with TBI remains to be determined. Furthermore, the widespread belief that all patients in a vegetative state are awake but not aware has been challenged. Incidental reports on functional MRI studies show that external stimuli can be processed in the human cortex of some vegetative patients, and that even spoken commands might elicit appropriate cortical responses that are indistinguishable from normal human responses.[55] The identification of biomarkers that can be used to predict these clinical outcomes would be invaluable to identify those patients likely to benefit from a rehabilitation regimen, to inform and counsel family members regarding the prognosis, and to track the level of consciousness in individuals who survive severe TBI. Many studies have reported on the association between predictors and outcome after TBI. High initial serum levels of S100B, UCH-L1 and NSE, GFAP, and SBDPs have all been linked to mortality 6 months later. Recently, Vos and colleagues showed a high specificity of GFAP in predicting death or unfavorable outcome at 6 months (0.93–0.95), with a false-positive rate for unfavorable outcome below

5%.[56] Furthermore, studies have shown that time course of biomarkers in non-survivors differed from survivors,[43,57] indicating that more careful monitoring of these biochemical markers might be appropriate.

Finally, in our recent work we have shown an association between a bio-marker level, i.e. MAP-2, in the chronic phase after TBI and the emergence to higher levels of cognitive function in individuals who survive severe TBI, suggesting a potential opportunity for a better understanding of the pathophysiology of the chronic phase after TBI.

12.5   Conclusions

At present, many unanswered questions remain in the area of brain bio-markers. The methodological and technological advances described here can lead to the construction of a coherent biomarker pipeline with a higher likelihood of success than past approaches. However, demonstration of clinical utility and compliance with regulatory requirements remain challenging, uncertain, and costly steps toward the commercialization of novel biochemical markers. Biomarkers have been proved to be useful in a routine clinical diagnostic setting but more detailed guidelines are needed on how biomarkers can be integrated into current diagnostic procedures. We should recognize that no single biomarker as well as no single treatment can be uniformly appropriate across the wide range of conditions within TBI, and this vision would support the search for a multimarker strategy that may provide a greatly expanded approach to the detection of brain injury and deep insights in its pathogenesis and pathomechanisms. This might be an important step towards individualized treatment approaches and, ultimately, improving outcome in subjects who have suffered from a severe TBI.

References

1. J. B. Fenn, M. Mann, C. K. Meng, S. F. Wong and C. M. Whitehouse, *Science*, 1989, **246**, 64.
2. J. D. Guingab, F. Kobeissy, M. Ratliff, Z. Zhang, K. W. Wang, Neurogenomics and Neuroproteomics Approaches of Studying Neural Injury, in *Essentials of Spinal Cord Injury*, ed. M. Fehlings, *et al.*, Thieme, Toronto, 2009.
3. P. Davidsson and M. Sjogren, *Dis. Markers*, 2005, **21**, 81.
4. J. E. Celis, P. Gromov, T. Cabezon, J. M. Moreira, N. Ambartsumian, K. Sandelin, F. Rank and I. Gromova, *Mol. Cell. Proteomics*, 2004, **3**, 327.
5. W. E. Haskins, F. H. Kobeissy, R. A. Wolper, A. K. Ottens, J. W. Kitlen, S. H. McClung, B. E. O'Steen, M. M. Chow, J. A. Pineda, N. D. Denslow, R. L. Hayes and K. K. Wang, *J. Neurotrauma*, 2005, **22**, 629.
6. B. K. Shin, H. Wang and S. Hanash, *J. Mammary Gland Biol. Neoplasia*, 2002, **7**, 407.
7. N. Denslow, M. E. Michel, M. D. Temple, C. Y. Hsu, K. Saatman and R. L. Hayes, *J. Neurotrauma*, 2003, **20**, 401.

8. T. Katano, T. Mabuchi, E. Okuda-Ashitaka, N. Inagaki, T. Kinumi and S. Ito, *Proteomics*, 2006, **6**, 6085.
9. Q. Ding, Z. Wu, Y. Guo, C. Zhao, Y. Jia, F. Kong, B. Chen, H. Wang, S. Xiong, H. Que, S. Jing and S. Liu, *Proteomics*, 2006, **6**, 505.
10. J. B. Redell, Y. Liu and P. K. Dash, *J. Neurosci. Res.*, 2009, **87**, 1435.
11. F. H. Kobeissy, S. Sadasivan, M. W. Oli, M. S. Gold and K. K. Wang, *Proteomics Clin. Appl.*, 2008, **2**, 1467.
12. K. K. Wang, S. F. Larner, G. Robinson and R. L. Hayes, *Curr. Opin. Neurol.*, 2006, **19**, 514.
13. A. K. Ottens, F. H. Kobeissy, B. F. Fuller, M. C. Liu, M. W. Oli, R. L. Hayes and K. K. Wang, *Prog. Brain Res.*, 2007, **161**, 401.
14. A. K. Ottens, F. H. Kobeissy and E. C. Golden, *Mass Spectrom. Rev.*, 2006, **25**, 380.
15. N. Nogoy, *Expert Rev. Proteomics*, 2007, **4**, 343.
16. N. Rifai, M. A. Gillette and S. A. Carr, *Nat. Biotechnol.*, 2006, **24**, 971.
17. J. A. Burgess, P. Lescuyer, A. Hainard, P. R. Burkhard, N. Turck, P. Michel, J. S. Rossier, F. Reymond, D. F. Hochstrasser and J. C. Sanchez, *J. Proteome Res.*, 2006, **5**, 1674.
18. D. Dumont, J. P. Noben, J. Raus, P. Stinissen and J. Robben, *Proteomics*, 2004, **4**, 2117.
19. B. N. Hammack, K. Y. Fung and S. W. Hunsucker, *Mult. Scler.*, 2004, **10**, 245.
20. S. I. Svetlov, Y. Xiang, M. W. Oli, D. P. Foley, G. Huang, R. L. Hayes, A. K. Ottens and K. K. Wang, *Biomarkers*, 2006, **11**, 355.
21. F. H. Kobeissy, A. K. Ottens, Z. Zhang, M. C. Liu, N. D. Denslow, J. R. Dave, F. C. Tortella, R. L. Hayes and K. K. Wang, *Mol. Cell. Proteomics*, 2006, **5**, 1887.
22. S. S. Chen, W. E. Haskins and A. K. Ottens, Bioinformatics for traumatic brain injury: Proteomic Data Mining, in *Data Mining in Biomedicine*, ed. P. M. Pardalos, V. L. Boginski and A. Vazacopoulos, Springer, 2007, pp. 1–26.
23. S. G. Grant, *Curr. Opin. Neurobiol.*, 2003, **13**, 577.
24. S. G. Grant and W. P. Blackstock, *J. Neurosci.*, 2001, **21**, 8315.
25. P. Beltrao, C. Kiel and L. Serrano, *Curr. Opin. Struct. Biol.*, 2007, **17**, 378.
26. K. K. Wang, A. K. Ottens, M. C. Liu, S. B. Lewis, C. Meegan, M. W. Oli, F. C. Tortella and R. L. Hayes, *Expert Rev. Proteomics*, 2005, **2**, 603.
27. R. Greenberg, *Am. J. Hosp. Pharm.*, 1976, **33**, 1308.
28. http://frwebgate.access.gpo.gov/cgi-bin/get-cfr.cgi?YEAR=current&TITLE=21&PART=807&SECTION=81&SUBPART=&TYPE=TEXT
29. http://frwebgate.access.gpo.gov/cgi-bin/get-cfr.cgi?YEAR=current&TITLE=21&PART=814&SECTION=1&SUBPART=&TYPE=TEXT
30. http://ecfr.gpoaccess.gov/cgi/t/text/textidx?c=ecfr&tpl=/ecfrbrowse/Title21/21cfr812_main_02.tpl 2006
31. http://www.fda.gov/RegulatoryInformation/Legislation/FederalFood-DrugandCosmeticActFDCAct/FDCActChapterVDrugsandDevices/ucm110188.htm accessed 2010-04-27 2009
32. http://www.fda.gov/MedicalDevices/DeviceRegulationandGuidance/IVDRegulatoryAssistance/ucm123682.htm#4b:accessed 2010-04-27 2010

33. R. Etzioni, N. Urban, S. Ramsey, M. McIntosh, S. Schwartz, B. Reid, J. Radich, G. Anderson and L. Hartwell, *Nat. Rev. Cancer*, 2003, **3**(4), 243.
34. F. Vitzthum, F. Behrens, N. L. Anderson and J. H. Shaw, *J. Proteome Res.*, 2006, **4**(4), 1086.
35. K. Blennow, *Nat. Med.*, 2010, **16**(11), 1218.
36. A. Raabe, C. Grolms, O. Sorge, M. Zimmermann and V. Seifert, *Neurosurgery*, 1999, **45**(3), 477.
37. B. Romner, T. Ingebrigtsen, P. Kongstad and S. E. Borgesen, *J. Neurotrauma*, 2000, **17**, 641.
38. B. Romner and T. Ingebrigtsen, *J. Neurosurg.*, 2001, **49**, 1490.
39. C. Woertgen, R. D. Rothoerl, C. Metz and A. Brawanski, *J. Trauma*, 1999, **47**, 1126.
40. R. Donato, *Int. J. Biochem. Cell. Biol.*, 2001, **33**(7), 637.
41. N. C. Ringger, B. E. O'Steen, J. G. Brabham, X. Silver, J. Pineda, K. K. Wang, R. L. Hayes and L. Papa, *J. Neurotrauma*, 2004, **21**, 1443.
42. K. J. Anderson, S. W. Scheff, K. M. Miller, K. N. Robert, L. K. Gilmer, C. Yang and G. Shaw, *J. Neurotrauma*, 2008, **25**, 1079.
43. S. Mondello, L. Akinyi, A. Buki, S. Robicsek, A. Gabrielli, J. Tepas, L. Papa, G. M. Brophy, F. C. Tortella, R. L. Hayes and K. K. Wang, *Neurosurgery*, 2011, [Epub ahead of print].
44. R. P. Berger, M. C. Pierce, S. R. Wisniewski, P. D. Adelson, R. S. Clark, R. A. Ruppel and P. M. Kochanek, *Pediatrics*, 2002, **109**, E31.
45. L. Papa, L. Akinyi, M. C. Liu, J. A. Pineda, J. J. Tepas, M. W. Oli, W. Zheng, G. Robinson, S. A. Robicsek, A. Gabrielli, S. C. Heaton, H. J. Hannay, J. A. Demery, G. M. Brophy, J. Layon, C. S. Robertson, R. L. Hayes and K. K. Wang, *Crit. Care Med.*, 2010, **38**, 138.
46. G. M. Brophy, S. Mondello, L. Papa, S. Robicsek, A. Gabrielli, J. Tepas, A. Buki, C. Robertson, F. C. Tortella, R. L. Hayes and K. K. Wang, *J. Neurotrauma*, 2011, **28**(6), 861.
47. S. Mondello, L. Papa, A. Buki, M. R. Bullock, E. Czeiter, F. C. Tortella, K. K. Wang and R. L. Hayes, *Crit Care*, 2011, **15**(3), R156.
48. J. A. Pineda, S. B. Lewis, A. B. Valadka, L. Papa, S. C. Heaton, J. A. Demery, M. C. Liu, J. M. Aikman, V. Akle, G. M. Brophy, J. J. Tepas, K. K. Wang, C. S. Robertson and R. L. Hayes, *J. Neurotrauma*, 2007, **24**, 354.
49. G. M. Brophy, J. A. Pineda, L. Papa, S. B. Lewis, A. B. Valadka, H. J. Hannay, S. C. Heaton, J. A. Demery, M. C. Liu, J. J. Tepas, A. Gabrielli, S. Robicsek, K. K. Wang, C. S. Robertson and R. L. Hayes, *J. Neurotrauma*, 2009, **26**, 471.
50. S. Mondello, S. Robicsek, A. Gabrielli, G. M. Brophy, L. Papa, J. J. Tepas, C. S. Robertson, A. Buki, D. Scharf, M. Jixiang, L. Akinyi, U. Muller, R. L. Hayes and K. K. Wang, *J. Neurotrauma*, 2010, **27**, 1203.
51. J. Ganesalingam, J. An, C. E. Shaw, G. Shaw, D. Lacomis and R. Bowser, *J. Neurochem.*, 2011, **117**(3), 528.
52. M. Douglas-Escobar, C. Yang, J. Bennett, J. Shuster, D. Theriaque, A. Leibovici, D. Kays, T. Zheng, C. Rossignol, G. Shaw and M. D. Weiss, *Pediatr. Res.*, 2010, **68**(6), 531.

53. L. E. Pelinka, A. Kroepfl, R. Schmidhammer, M. Krenn, W. Buchinger, H. Redl and A. Raabe, *J. Trauma*, 2004, **57**, 1006.
54. D. M. Stein, A. Lindell, K. R. Murdock, J. A. Kufera, J. Menaker, K. Keledjian, G. V. Bochicchio, B. Aarabi and T. M. Scalea, *J. Trauma*, 2011, **70**(5), 1096.
55. A. M. Owen and M. R. Coleman, *Nat. Rev. Neurosci.*, 2008, 235.
56. P. E. Vos, K. J. Lamers, J. C. Hendriks, M. van Haaren, T. Beems, C. Zimmerman, W. van Geel, H. de Reus, J. Biert and M. M. Verbeek, *Neurology*, 2004, **62**, 1303.
57. R. P. Berger, M. C. Bazaco, A. K. Wagner, P. M. Kochanek and A. Fabio, *Dev. Neurosci.*, 20, **32**(5–6), 396.

CHAPTER 13

Future Trends in Biomarker Immunoassay Development

SVETLANA A. DAMBINOVA*[a] AND RONALD L. HAYES[b]

[a] Kennesaw State University, Brain Biomarkers Laboratory, 1000 Chastain Road, Kennesaw, GA 30144, USA; [b] Banyan Biomarkers, 13400 Progress Blvd., Alachua, FL 32615, USA
*Email: sdambino@kennesaw.edu

13.1 Introduction

Clinical questions regarding an individual's condition after mild TBI differ from research questions concerning biomarkers.[1] Despite the discovery of thousands of biomarkers, which has led to research on hundreds of those potentially applicable to patient care, only a small number have been validated and shown evidence for utility in diagnostic assessment of disease.[2] The evaluation of specific biomarkers for assessment of TBI and other neurological conditions is in its infancy.

Requirements for TBI testing include (i) the exact clinical indication for the assay; (ii) analyte detection and assay performance characteristics; (iii) potential utility for guiding treatment selection and monitoring, and (iv) prognostic outcome. A diagnostic blood test that can detect specific biomarkers for TBI – preferably, as an immunoassay – should be reliable and user-friendly.[2]

The development of an immunoassay is remarkably complex, calling for a demanding multidisciplinary approach, knowledge, and skills. Principal objectives of blood assay development include identification of biomarkers directly involved in the pathophysiology of disease; development of rare reagents, including key antigen and corresponding polyclonal or monoclonal

RSC Drug Discovery Series No. 24
Biomarkers for Traumatic Brain Injury
Edited by Svetlana A. Dambinova, Ronald L. Hayes and Kevin K. W. Wang
© Royal Society of Chemistry 2012
Published by the Royal Society of Chemistry, www.rsc.org

antibodies; selection of format and technology; optimization of assay conditions; and assay validation and feasibility studies. To meet these objectives, assay designers need to balance many factors, including those imposed by quality, cost, and time.[3–5]

Evaluation of clinical efficacy and safety is required before an immunoassay can be incorporated into patient care. Country-specific regulatory guidelines are responsible for protecting and promoting public health. Despite individual differences, all countries coordinate multicenter clinical trials and regulate the clearance process for immunoassay dissemination for clinical laboratory testing based on demonstrated results among the populations studied.

In this chapter, we provide information concerning the inter-relationship between assay development, clinical indications, regulatory restrictions, and future trends in drug/test co-development, or "companion diagnostics," for TBI biomarkers. The detailed strategy of biomarker assay development, use of diagnostic biomarkers for drug development, and adaptation of clinical laboratories to personalized medicine may be found in other publications.[2–8]

13.2 TBI Biomarkers: Assay Requirements

According to the U.S. Food and Drug Administration (FDA), an ideal biomarker must be specifically associated with a particular disease or disease state.[2] A blood assay designed to detect specific biomarkers should be able to differentiate between similar physiological conditions, resulting in rapid, simple, accurate, and inexpensive detection of the relevant biomarker.

13.2.1 Clinical Indications

The treatment decision-making process for TBI requires evidence from clinical observations and neuropsychological testing for mild TBI and from neuroimaging, which currently represents the best modality for diagnosis of moderate to severe TBI. Addition of a blood assay that can reliably detect brain biomarker(s) for TBI could improve the diagnostic certainty of distinct pathological conditions. Therefore, the most important clinical indication(s) for an immunoassay are those that reflect primary (acute) or secondary (chronic) events as well as CNS injury severity.

Clinical indication labeling for a blood test is a simple and direct way to communicate with clinical laboratories and clinicians about the decision-making process.[9,10] Use of such tests has become routine in many clinical practices (e.g. diabetes, oncology, and cardiovascular and infectious diseases) and is acceptable to the medical community. However, for neurology and neurosurgery, such tests are currently available only as research tools.

Medical needs that should be addressed with respect to diagnosis of mild TBI sequelae include the following clinical indications for acute events: (i) differential diagnosis of acute primary injury versus normal population;

(ii) primary versus multiple concussion; (iii) mild TBI versus mimics; and (iv) mild TBI versus incomplete spinal cord injury. Clinical indications for secondary injury (chronic conditions) that would result in better patient care include: (i) diagnosis of post-traumatic stress disorder (PTSD); (ii) hemorrhage versus ischemia; (iii) stroke risk; and (iv) diagnosis of post-traumatic epilepsy (PTE).

To fulfill these unmet clinical indications, a number of brain/spinal cord biomarkers that show reliable test results in different settings should be considered for use in a single biomarker assay and/or multiple panels of bio-markers. The best use for a single biomarker assay would be in the emergency department, by paramedics, and in the field as a point-of-care device,[11] while multiple panels (automatic analyzers) would be useful in certified clinical laboratories, including those in hospitals.[12,13]

A single biomarker assay should be capable of supplying the end user with a "yes" or "no" response, as is done currently with troponin and brain natriuretic peptide in cardiovascular diagnostics. In contrast, a multiple panel of biomarkers may assess TBI severity (i.e. mild, moderate, or severe) or TBI sequelae (i.e. hemorrhage, ischemia, PTSD, or PTE). Research data are needed to confirm whether early diagnosis of TBI/spinal cord injury and consequences of CNS damage can be predicted with the use of biomarker assays.

13.2.2   Preliminary Assay Performance Characteristics

Before a diagnostic biomarker can enter clinical practice, analyte values detected in patients' biological fluids need to be correlated with disease (i.e. TBI) or "normal" status (i.e. non-TBI). The clinical utility characteristics of the analyte being measured by the assay should be explored in well-defined population cohorts and cut-off values for each group estimated.

Cut-offs values represent the threshold between TBI and non-TBI cohorts. In addition to healthy individuals without injuries, a non-TBI cohort may comprise persons with symptoms mimicking TBI (e.g. migraine, syncope, vertigo, sepsis, or drug intoxication), as well as those with pre-existing conditions (e.g. hypertension, diabetes, atherosclerosis, or cardiovascular disease).

To establish cut-off values, the major test characteristics – sensitivity and specificity – should be determined.[14] As a variable, sensitivity measures the percentage of TBI patients who are correctly identified as having the abnormal condition. Specificity, another variable, measures the percentage of healthy people who are correctly identified as not having CNS injury. The following example outlines performance characteristics calculations.[1] Two cohorts form the basis for consideration and estimation: cohort 1 includes those with no evidence of TBI, while cohort 2 includes persons with TBI. Determinations of the analyte are performed with all samples from both cohorts and the values ranked. A test is valid if a clear separation of values for the two cohorts is found and the likelihood ratio is above 5 (Table 13.1) and if a positive test indicates the disorder is present (high positive predictive value).[14]

Table 13.1 The interpretation of positive likelihood ratios (LR) resulting from immunoassay testing.

LR	Interpretation
>10	Large and often conclusive increase in the likelihood of disease
5–10	Moderate increase in the likelihood of disease
2–5	Small increase in the likelihood of disease
1–2	Minimal increase in the likelihood of disease
1	No change in the likelihood of disease

The receiver operating characteristic (ROC) curve should be plotted to calculate the cut-off value for optimal sensitivity, or true positive rate, versus optimal specificity, or false positive rate. The best sensitivity is usually presented within a range of 85%–95%, whereas specificity rarely reaches 98%.[15]

Establishing preliminary performance characteristics for several different patient cohorts, including disorders mimicking TBI, will help inform immunoassay limitations in various physiological and pathological conditions.

13.2.3 Peculiarities of Brain Biomarkers for Immunoassays

Development of immunoassay prototypes based on brain biomarkers for rapid assessment of brain damage require the preparation of rare reagents that contain specific antigens and corresponding polyclonal or monoclonal antibodies. Synthetic antigens are usually small molecular weight protein/peptides that may be hydrophobic, have a tendency to aggregation or oligomerization, and oxidize during storage. To preserve peptide immunogenic activity and stability, chemical modification is a necessity.[16]

High-affinity polyclonal and/or monoclonal antibodies are used as capture and detection reagents. These antibodies are also usually utilized as calibrators and standards for indirect antibody testing. Use of hen yolk immunoglobulins (IgY) can help prevent cross-reactivity observed with other IgG, even those with low or absent species, strains, and other immunogens.[17]

13.3 Diagnostic Challenges in Development of Immunoassays for TBI

Several factors play a significant role in immunoassay development, including selection of assay architecture (including specimen type, assay type, and solid matrix); selection of immunoassay format (manual, automatic, or point-of care); and bench testing for assay prototype.

13.3.1 Specimen Choice

The selection of assay specimen type, sample collection standardization, and storage conditions should be specified to minimize immunoassay variability. The best choice of matrix for a brain biomarker assay is plasma (antigen assay) and/or serum (antibody assay). For the peptide assay, EDTA plasma often is the only suitable sample.[18] Serum contains a number of serine and other proteases that are activated during coagulation (from fibrinogen to fibrin) and cause depletion of free peptide. Plasma quality plays an important role; use of hemolyzed plasma is not recommended.[18] Hemolyzation occurs when a blood sample has been drawn and left unattended at room temperature. For every hour that hemolyzed plasma is stored at room temperature, the free peptide level diminishes by 20%–25% (unpublished data). Cerebrospinal fluid is seldom used, due to the invasive nature of intraventricular and lumbar puncture. However, for severe TBI and spinal cord injury cases it could be only choice to predict poor prognosis.[6]

13.3.2   Choice of Immunoassay Architecture

Heterogeneous and homogeneous immunoassays are categorized based on how differentiation between reactive and non-reactive components is accomplished. Both use immobilized antigen or antibodies on a solid surface and can be further characterized as competitive (reagent limited), or non-competitive (reagent excess) assays (Figure 13.1). A number of different solid matrices have been developed utilizing various plastic surfaces and spherical micro- or nanoparticles.

The rapidly growing and expanding use of nanoparticles in immunoassay development is predicated by the need for more sensitive detection of protein-based biomarkers, especially for earlier TBI diagnosis.[19,20] The availability of magnetic mono-sized polymer-coated particles has helped overcome disadvantages of plastic surfaces by increasing assay sensitivity threefold to six-fold and reducing turnaround time.

13.3.3   Immunoassay Format

Several standard formats for immunoassays are currently in use in different clinical settings: primary care, outpatient/inpatient, emergency department, and bedside. These include the manual enzyme-linked immunoassay (ELISA) used primarily by reference laboratories; automatic (immunoassay analyzers) adjusted to high volume clinical laboratories; and point-of-care (POC) utilized by primary care, in the emergency department, and in the home (e.g. glucose, cholesterol, ketone, and pregnancy).

An ELISA is the most popular format for development of novel biomarker immunoassays[7] and used mainly to detect the presence of an antibody or an antigen in a biological sample. This format is attractive as an assay due to its sufficient sensitivity and efficiency at examining rare reagents and is safe and inexpensive. Drawbacks include that a standard ELISA procedure is characterized by a relatively prolonged turnaround time (1.5–3.0 hours); difficulties with deviations in well surface (>10%); and chromophore alterations after quenching enzyme activity (Table 13.2). To avoid these limitations, the use of nanoparticles and chemo- or fluoro-luminescence reaction detections are recommended.[20] This approach has been applied to automatic immunoassays with high throughput and a multiple menu of biomarkers (Table 13.2).

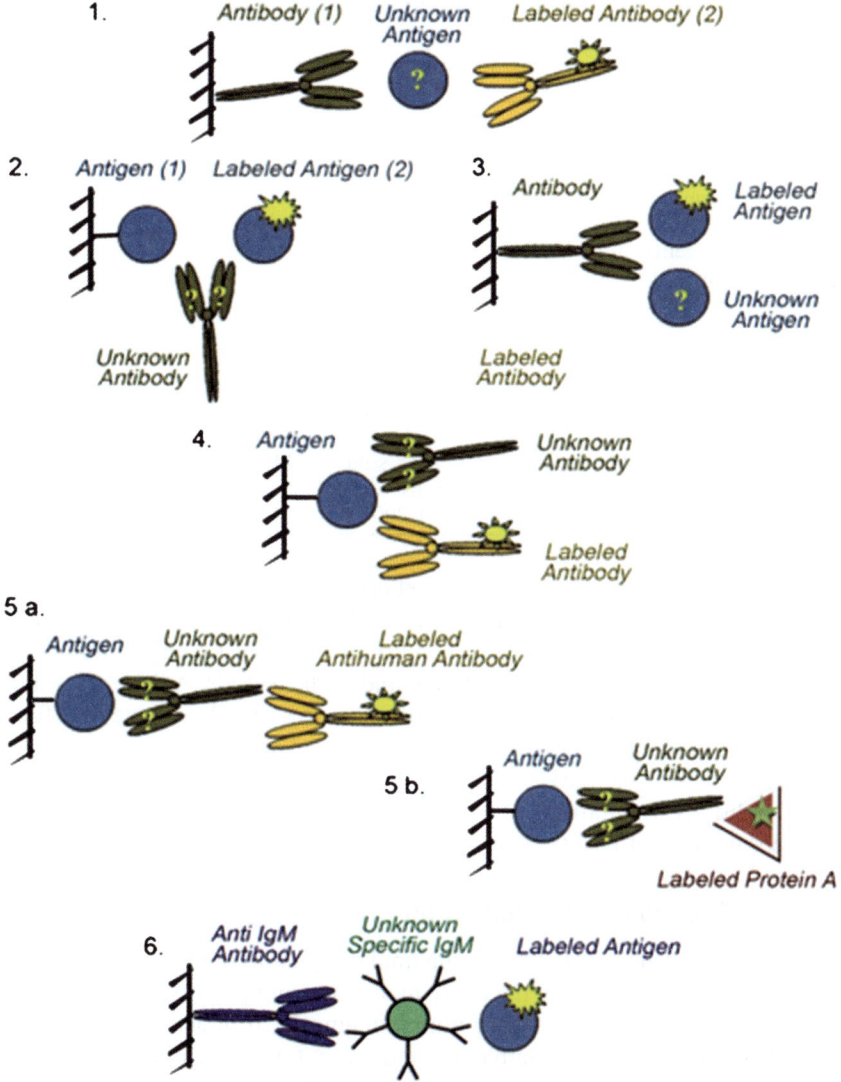

Figure 13.1 Methodologies in immunochemistry. 1: sandwich assay when double antibodies determine unknown antigen; 2: sandwich assay when double antigen detects unknown antibodies; 3: competitive assay when unknown antigen competes for attached to solid surface antibodies with known concentration of labeled antigen; 4: competitive or serological assay when unknown antibodies compete for attached to solid surface antigen with known concentration of labeled antibodies; 5: "serological assay" when unknown antibodies react to attached to solid surface antigen and detected by labeled anti-human antibodies (a) or labeled protein A (b); 6: IgM-specific assay when the unknown specific IgM antibodies react with anti-IgM antibodies attached to solid phase and reaction is revealed by labeled antigen.

Table 13.2 Manual and automatic analyzer immunoassay formats: advantages and medical needs.

Parameter	Assay Manual (M)	Automatic (A)	Advantage	Medical needs
User friendly	No	Yes	A	Yes
Turnaround time	3 hours (90 samples)	1 hours (1200 samples)	A	Should be decreased (<10 min)
Number of reagents	5–10	5	A	1
Sample withdrawal, preparation, storage	0.5–1 h	<0.5 h	A	Bedside & on field detection
Professional skills	High complicity	High complicity	—	Non/home testing
Results interpretation	Manual	Automatic	A	Immediate
Waste (turnaround)	Plastic, solutions <1 L	Plastic, solutions <10 L	M	Minimal
Costs of 1 assay	<$100	$200–300	M	<$10

From a practical perspective, for effective assay use a brain biomarker immunoassay should be translated into a POC device that targets acute and chronic presentation of TBI in field conditions (e.g. sports, combat, and disaster areas), and in the emergency department and at the bedside. A POC assay allows simplification of sample preparation and minimizes sample volume (e.g. for use in pediatric patients), obtains a clear "yes"/"no" result for both single and multiple panels of biomarkers, and can be used to screen patients with chronic TBI conditions. Limitations of a POC assay may include accuracy, reliability, and repeatability of results, as has been found with POC glucose testing.[21]

13.3.4 Technology Optimization

Technology optimization includes the final design and testing. The goal of this phase is to build and verify the assay manufacturing prototype. By the end of this development step, the final immunodiagnostic product performance and format should be nearing completion and ready for pilot clinical trials. At this point, the manufacturer should meet with an FDA regulatory consultant to discuss the clinical tests required for regulatory submission; i.e. 510(k) or premarket approval (PMA) application in the United States. European Union regulatory requirements allows clinical use of an immunoassay with a specific clinical indication once it attains a CE (Conformité Européene) mark. Both FDA and CE clearances are accepted in the majority of countries worldwide, with slight adjustments required by local regulatory committees.

13.4 Safety Regulatory Stages for Immunoassays

In the U.S., the FDA approval process for diagnostic devices comprise three major phases: analytical, clinical phase 1, and clinical phase 2 (Figure 13.2).

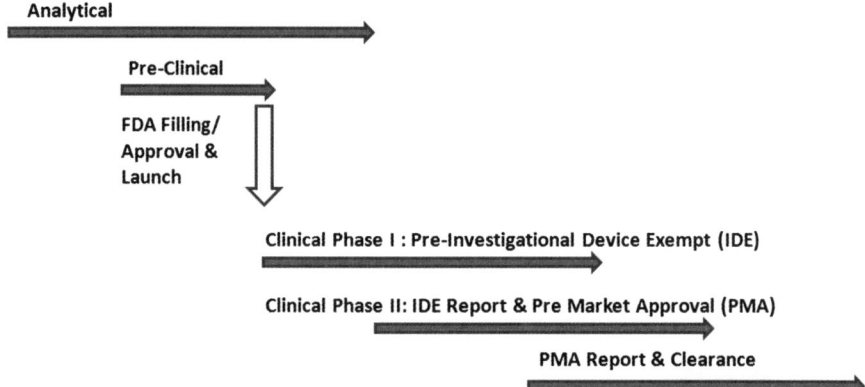

Figure 13.2 FDA requirements for diagnostic devices approval process sections.

The analytical phase is exploratory and includes immunoassay bench testing, creating the investigative research protocol for the initial clinical study, and a pre-Investigative Device Exemption (pre-IDE) from the FDA.[2,9] The pre-IDE allows the device to be used in a clinical study to collect safety and effectiveness data to support a PMA or 510(k) submission. Clinical phase I requires organization of multicenter clinical trials and data collection and processing. Clinical phase 2 is devoted to preparing the IDE report to gain PMA. Since TBI biomarkers are mostly novel and no comparative "gold standard" yet exists, biomarker immunoassays for TBI are treated as "a significant risk device" during the decision-making process and as such are required to take the PMA regulatory track (Figure 13.3). Regardless of the existence of a predicate device (i.e. one already cleared), 510(k) "de novo" (or "fast track") approval process is usually not considered by the FDA for stroke, TBI, spinal cord injury, and other acute neurological conditions requiring emergent treatment and diagnostic devices for these disorders are generally considered as Class III (i.e. significant patient risk).

In addition to clearance of devices based on novel biomarkers, the FDA recently introduced a procedure for a novel biomarker approval process. If a biomarker can be detected using Class I or II devices (i.e. there is low or moderate patient risk) and there is a high clinical need, a novel biomarker may be approved using the 510(k) de novo process.

13.4.1 Analytical Phase

Prior to clinical phase 1 testing, the assay is returned to the laboratory to test its reproducibility and accuracy using assay components and reagents (for example, synthetic peptides, antibodies, and conjugates for an immunoassay). The process in this analytical phase (bench testing) includes trueness, accuracy, repeatability, and reproducibility as well as determination of linearity and limits of detection and quantification.[9] Biomarker immunoassay reference

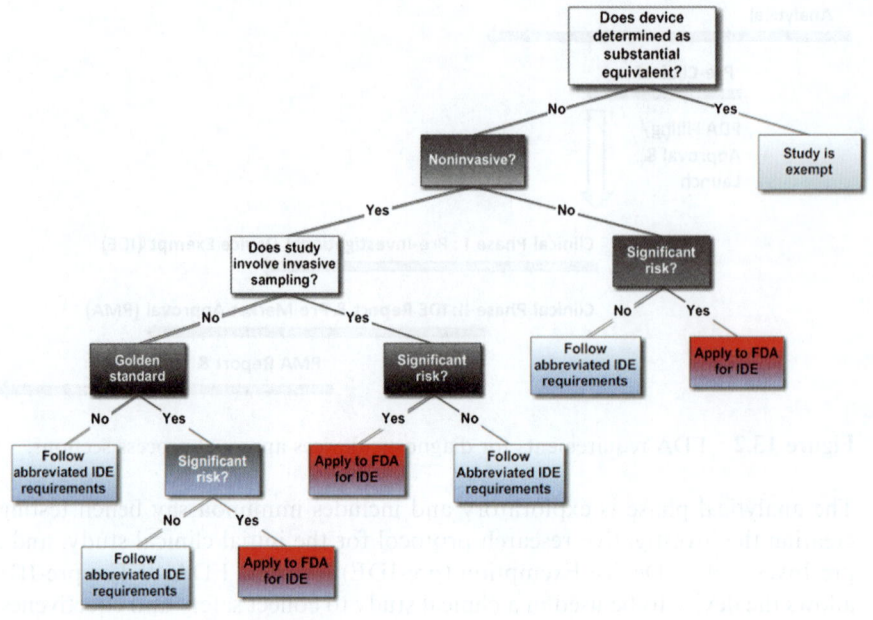

Figure 13.3   FDA regulatory stages for IVD.

intervals for a healthy population with respect to gender, age, and racial differences are then established.

   In addition, preclinical characterization of variability comprising physiological and non-physiological components is required and should be reflected in the protocol submitted to an Institutional Review Board (IRB) prior to enrolling patients in a clinical trial. Physiological components comprise within- and between-subject variations, effects of gender, age, ethnicity, pre-existing conditions, and medication use; while the non-physiological components addresses specimen type, patient preparation, and procedures such as CT or MRI. Delineating immunoassay advantages and limitations resulting from this phase will minimize variability and reduce any unknown factors prior to initiation of clinical trials.

13.4.2   Clinical Phase I and II

Once a manufacturing prototype has been built and verified, actual clinical testing of the assay can begin.[10] Clinical validation represents the most costly and time-consuming process in novel blood immunoassay development and, since the goal is to obtain regulatory approval, results using the assay must meet several criteria, including high sensitivity/specificity and clinical feasibility relative to the objective standard which, in the case of TBI, is neuro-imaging. Medical Centers of Excellence that are experienced in FDA trials should conduct these studies. In addition, the services of third-party

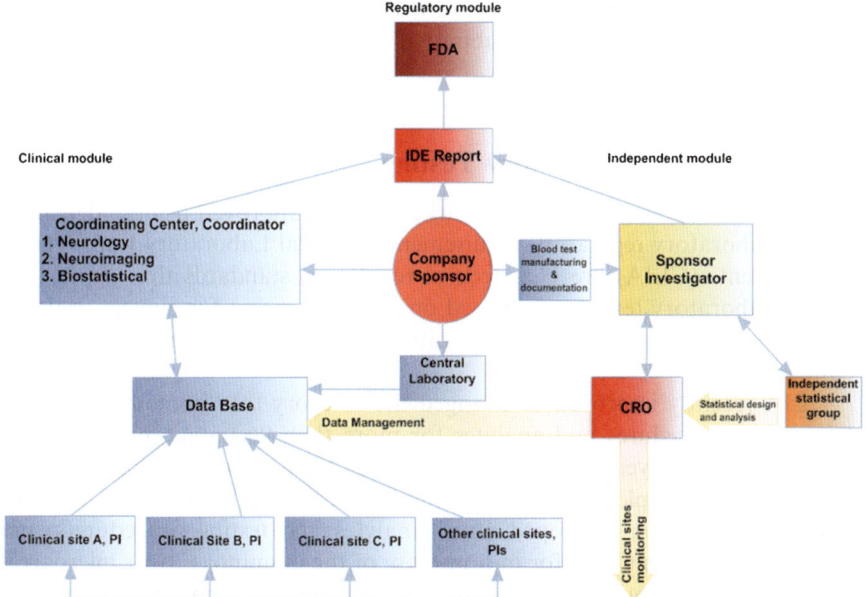

Figure 13.4 Clinical trials chart for IDE application.

professionals, a sponsor investigator, research coordinators, biostatisticians, and a data management center are often contracted to handle data collection and regulatory submission. The central laboratory should be certified for clinical trials and may either be contracted or connected to the clinical trial sponsor (Figure 13.4). Clinical trials should be double-blind, prospective, multicenter, and regulated by an Investigative Device Protocol that has been approved by each clinical site and has met requirements for an FDA pre-IDE. A number of pre-IDEs may be required to be created based on different clinical indications (see Section 13.2.1).

For example, a protocol or pre-IDE for an acute TBI biomarker assay trial should have the clinical indication specified and refer to a specific population cohort. A pre-IDE might comprise two steps: the first step addresses confirmation of preliminary immunoassay cut-off(s) for specific time points to demonstrate peak biomarker efficacy; for example, 0–3 hours, 3–6 hours, 6–12 hours, or 12–24 hours. Then, when the optimal cut-off is set for a determined time window, a group of patients has to be enrolled to assess immunoassay performance characteristics (e.g. sensitivity, specificity, positive and negative predictive values, and likelihood ratios).

Clinical trial information, results of clinical observations, radiological findings, neuropsychological testing results, and clinical laboratory data presented in case report forms (CRFs) are filed in a database with restricted access. Interpretation of database results by an independent biostatistical group is summarized in a report that is presented to the sponsor. The contract research organization (CRO), in collaboration with the sponsor investigator, then prepares an IDE report, which is submitted to a country's regulatory committee for pre-market approval.

13.4.3 Clinical Laboratory Testing

The testing process for a new biomarker may be approved using independent clinical laboratory regulatory requirements. Clinical Laboratory Improvement Amendments (CLIA) are U.S. federal regulatory standards that apply to all clinical laboratory testing performed on humans, except for clinical trials and basic research.[22] Analogous committees function in Europe and other countries.

CLIA recently improved or changed the category for "home brew" testing laboratory procedures and analytes for more than 96 unique protein targets and 211 genes assayed in plasma.[23] Under CLIA regulations, tests are classified according to complexity: "high," "moderate," and "waived." A reference laboratory that has developed novel analytes and testing procedure(s) may submit a request to CLIA for in-house use of the immunoassay or microarray. These procedures primarily concern drug screening, chronic conditions, and rare genetic disorders. Only procedures that predict risk of consequences of CNS injury (e.g. stroke, PTE, PTSD) may be CLIA certified.

CLIA-waived tests are those that are defined as simple and accurate, pose no risk to the patient, and have been FDA cleared for home use. Historically, they were presented as POC devices serving acute care; however, POC devices are now being used for monitoring chronic conditions such as diabetes and heart failure as well as anticoagulation treatment with warfarin. POC devices can assist in more rapid decision-making in different clinical settings (Table 13.3) and offer promise as diagnostics for neurological and neurosurgical conditions.

Table 13.3 Opportunities for use of point-of-care (POC) TBI tests.

Setting	Application	Benefit
Paramedic vehicle	Prehospital testing of severity	Faster triage in the ED and earlier intervention
Emergency room	Testing for rapid triage of cerebrovascular or post-traumatic complications	Reduce time in the ED and allow emergent treatment
Surgery room	Presurgical assessment of preexisting conditions and monitoring brain circulation parameters	Reduce postoperative care, adjust procedure, and improve outcome
Primary care office	Regular physical examination	Early detection of risk of complications and co-morbid disorders
Home	Monitoring of chronic post-traumatic conditions	Better awareness of and motivation to manage condition, and assistance to physician in patient managements

ED = emergency department

5. D. Wild (ed.), *The Immunoassay Handbook*, 3rd edn, Elsevier Science Publishing, New York, 2005.
6. E. J. Thompson, *Proteins of the Cerebrospinal Fluids*, Elsevier, London, 2005, p. 332.
7. J. R. Crowther, *The ELISA Guidebook*, Humana Press, Totowa, 2001, p. 421.
8. L. J. Kriska and G. H. G. Thorpe, in *Point-of-Care Testing*, ed. C. P. Price, A. St. John and L. L. Kriska, AACC Press, Washington, 2010, p. 27.
9. N. Rifai and R. E. Gerszten, *Clin. Chem.*, 2006, **52**, 1635.
10. G. L. Hortin, S. A. Carr and N. L. Anderson, *Clin. Chem.*, 2010, **56**, 149.
11. K. Silvester and C. P. Price, in *Point-of-Care Testing*, ed. C. P. Price, A. St. John and L. L. Kriska, AACC Press, Washington, 2010, p. 157.
12. http://www.ivdtechnology.com/article/new-brain-marker-laboratory-assessment-tia-and-stroke (last accessed September 2011).
13. S. Vanni, G. Polidori, G. Pepe, M. Chiarlone, A. Albani, A. Pagnanelli and S. Grifoni, *J. Emerg. Med.*, 2011, **40**(5), 499.
14. T. Greenhalgh, *Br. Med. J.*, 1997, **315**, 540.
15. S. A. Dambinova, G. A. Khounteev, G. A. Izykenova, I. G. Zavolokov, A. Y. Ilyukhina and A. A. Skoromets, *Clin. Chem.*, 2003, **49**, 1752.
16. M. H. V. Van Regenmortel, in *Synthetic Peptides as Antigens*, ed. M. H. V. Van Regenmortel and S. Muller, Elsevier, Amsterdam, 1999, p. 1.
17. R. Schade, E. G. Calzado, R. Sarmiento, P. A. Chacana, J. Porankiewicz-Asplund and H. R. Terzolo, *Altern. Lab. Anim.*, 2005, **33**(2), 129.
18. W. G. Guder, S. Narayanan, H. Wisser and B. Zawta, *Diagnostic Samples: From the Patient to the Laboratory*, 4th edn., Wiley-Blackwell, Darmstadt, 2009, p. 113.
19. A. S. Jaffe, *Clin. Chim. Acta*, 2007, **381**, 9.
20. J. Carney, H. Braven, J. Sea and E. Whitworth, *IVD Technology*, 2006, **3**, 41.
21. I. B. Hirsch, B. W. Bode and B. P. Childs, *Diabetes Technol. Ther.*, 2008, **10**, 419.
22. J. O. Westgard, S. S. Ehrmeyer and T. P. Darcy, *CLIA Final Rules for Quality Systems*, Westgard QC, Madison, 2004, p. 224.
23. N. L. Anderson, *Clin. Chem.*, 2010, **56**(2), 177.
24. Centers for Disease Control and Prevention, *M.M.W.R. Weekly*, 2007, **56**(19), 469.
25. http://www.drugs.com/newdrugs/fda-approves-zelboraf-companion-diagnostic-test-late-stage-skin-cancer-2814.html (last accessed September 2011).
26. S. A. Dambinova and G. A. Izykenova, *Ann. N.Y. Acad. Sci.*, 2002, **965**, 497.
27. http://www.cancer.gov/dictionary?CdrID=561717 (last accessed September 2011).
28. S. A. Dambinova, *NMDAR Biomarkers for Diagnosing and Treating Cerebral Ischemia*, PCT/US2007/087278, December 12, 2007.
29. C. Foerch, I. Curdt, B. Yan, F. Dvorak, M. Hermans, J. Berkefeld, A. Raabe, T. Neumann-Haefelin, H. Steinmetz and M. Sitzer, *J. Neurol. Neurosurg. Psychiatry*, 2006, **77**, 181.
30. Biomarkers Definitions Working Group, *Clin. Pharmacol. Ther.*, 2001, **69**, 89.

Subject Index

Locators in **bold** refer to figures/diagrams